Surviving Democracy

Is democracy, in its neoliberalized form, responsible in part for bringing us to the brink of self-destruction and the policy inertia that is doing away with our chances of survival? *Surviving Democracy* probes the way that democracy became neoliberalized and the role that neoliberalized democracy plays in our dealings with—causing, understanding, denying, and hopefully, mitigating—climate change.

Defining neoliberalism as the art of exclusion through inclusion, Chien-Yi Lu treats climate change as collateral damage of the neoliberal order established to ensure upward power and wealth redistribution. Highlighting the role that money played in the "free" competition of ideas between John Maynard Keynes and Friedrich Hayek, she investigates the resulting global structure, wherein the wealthy and powerful sit above the market and democracy, and the way that this structure fundamentally contradicts honest climate mitigation. Central to the structure is neoliberal elites' leveraging of the fluid relationship between the market and the state. Merging citizen power with consumer and investor powers is therefore imperative to the success of climate action. While expediting the bursting of the carbon bubble is an obvious answer, it is the discussion of the meat bubble that brings the book full circle, linking our survival to neoliberalism, inclusion, and democracy.

Chien-Yi Lu is an Associate Research Fellow at the Institute of European and American Studies of Academia Sinica in Taipei, Taiwan. Her research focuses on how democracy has transmuted as regional integration has deepened and supranational institutions have gained more power, and how democracy, in its neoliberalized form, has intervened in our ability to recognize, curb, and cope with climate change.

Environmental Politics

Edited by Steve Vanderheiden, University of Colorado at Boulder.

Over recent years environmental politics has moved from a peripheral interest to a central concern within the discipline of politics. This series aims to reinforce this trend through the publication of books that investigate the nature of contemporary environmental politics and show the centrality of environmental politics to the study of politics per se. The series understands politics in a broad sense and books will focus on mainstream issues such as the policy process and new social movements as well as emerging areas such as cultural politics and political economy. Books in the series will analyse contemporary political practices with regards to the environment and/or explore possible future directions for the 'greening' of contemporary politics. The series will be of interest not only to academics and students working in the environmental field, but will also demand to be read within the broader discipline.

The series consists of two strands:

Environmental Politics addresses the needs of students and teachers, and the titles will be published in paperback and hardback. Titles include:

EU Enlargement and the Environment
Institutional change and environmental policy in Central and Eastern Europe
Edited by JoAnn Carmin and Stacy D. VanDeveer

The Crisis of Global Environmental Governance
Towards a new political economy of sustainability
Edited by Jacob Park, Ken Conca and Matthias Finger

Routledge Research in Environmental Politics presents innovative new research intended for high-level specialist readership. These titles are published in hardback only and include:

Organizing for Policy Influence
Comparing Parties, Interest Groups, and Direct Action
Benjamin Farrer

Surviving Democracy
Mitigating Climate Change in a Neoliberalized World
Chien-Yi Lu

Surviving Democracy
Mitigating Climate Change in A Neoliberalized World

Chien-Yi Lu

Routledge
Taylor & Francis Group

LONDON AND NEW YORK

First published 2020 by Routledge

2 Park Square, Milton Park, Abingdon, Oxon OX14 4RN

605 Third Avenue, New York, NY 10017

Routledge is an imprint of the Taylor & Francis Group, an informa business

First issued in paperback 2021

Publisher's Note

The publisher has gone to great lengths to ensure the quality of this reprint but
points out that some imperfections in the original copies may be apparent.

Library of Congress Cataloging-in-Publication Data
A catalog record for this title has been requested

ISBN: 978-1-138-06190-3 (hbk)
ISBN: 978-1-03-217268-2 (pbk)
DOI: 10.4324/9781315162010

Typeset in Times New Roman
by Taylor & Francis Books

To mom and dad and in memory of grandma

Contents

Foreword

In 1988 the American climate modeler James Hansen testified in the U.S. Congress that anthropogenic climate change was under way. In the same year, the Intergovernmental Panel on Climate Change (IPCC) was assembled to advise governments what to do about this existential challenge. The fact that the IPCC was created as a joint venture between a scientific organization—the World Meteorological Organization (WMO)—and a political one—the United Nations (UN)—tells us that Hansen was not out on a limb; scientists had for some time recognized the potential severity of this challenge.

Post-war advances in theory and data gathering had made it clear in the 1950s that an increased concentration in atmospheric greenhouse gases (GHG) in the atmosphere, from burning fossil fuels, had the potential to change the planetary climate dramatically. Although scientists at that time did not necessarily know that the changes would all be adverse, they thought that many would be, particularly sea-level rise, which could drown coastal ports and cities around the globe. In the 1960s scientists began to try to alert governments and stimulate attention to the question. When the Clean Air Act was debated in the United States Congress in late 1969/early 1970, the subject of the "greenhouse effect," and its potential to cause global environment damage was one point of discussion. The private sector was also aware of the issue: historical documents show that various oil and gas companies, the American Petroleum Institute, and the Electric Power Research Institute already knew in the 1950s that GHG, from burning fossil fuels, could change the global climate.[1] In 1958 Frank Capra, one of Hollywood's most celebrated film-makers, produced an episode for an American television series about how human activities, including pollution, could change the weather.[2]

These activities and concerns were not limited to the United States. In 1972, when world leaders gathered in Stockholm for the United Nations Conference on the Human Environment—the first UN conference to address what we now call sustainability, climate change was a topic of conversation and a subject in the conference report.[3] Seven years later, climate was the focus when 350 specialists from 53 countries, 24 international organizations, and a wide range of disciplines including agriculture, water resources, fisheries, energy, environment, ecology, biology, medicine, sociology, and economics, gathered in

Geneva for the world's first international conference on "climate and mankind".[4] The conference report called on the nations of the world to "foresee and prevent potential man-made changes in climate that might be adverse to the well-being of humanity." Following this meeting, the WMO established the World Climate Programme, an umbrella for a diverse set of scientific and educational activities. And in 1985 the WMO, together with the UN Environment Programme and the International Council of Scientific Unions, convened in Villoch, Austria, for what in retrospect could be considered the first scientific assessment of the impact on the climate of GHG. The report of this meeting predicted that, unless steps were taken to control GHG emissions, global temperatures would rise dramatically, yielding a wide range of social, political, and economic impacts. Villoch paved the way for the creation of the IPCC and for the 1992 United Nations Framework Convention on Climate Change (UNFCCC), which committed its nearly 200 signatories to global action to prevent "dangerous anthropogenic interference" with the climate system.

Since that time, there have been hundreds of scientific reports and assessments and tens of thousands of peer-reviewed scientific papers on the issue. Laws have been passed, statements of commitment made, and resolutions galore. There have been countless conferences, reports, and opinion pieces. In 2019 parties to the UNFCCC met for the 25th time. Furthermore, in the last few years, universities, foundations, churches, and individuals have divested from fossil fuels, and citizens across the globe have taken to the streets.

However, despite all this work by scientists, political leaders, activists, and citizens, climate change has not been stopped. It has not even been slowed. At the start of the industrial revolution atmospheric carbon dioxide (CO_2) is estimated to have stood at about 280 parts per million (ppm); it has now passed 400 ppm, just as scientists predicted that it would. The last time that CO_2 concentrations were this high was more than 3 million years ago. Scientific data suggest that the global temperature at that time was 3°–4°C higher than it is today. The sea level was 15–25 meters higher, meaning that all of today's coastal cities, had they existed then, would have been under water.[5] Moreover, of course, when the other GHG that we have added to the atmosphere are taken into account, such as methane and certain short-lived fluorinated gases, the problem is even worse.[6] No matter how you interrogate the pronoun, *we* have failed to stop disruptive climate change. Scientists have failed, politicians have failed, activists have failed.

Chien-Yi Lu wants us to take this situation seriously and answer the question: Why? Why have we failed to act on climate change? Why have we failed to take steps to protect ourselves from this titanic threat? How *could* we have been so stupid?

Various authors, including me, have addressed the issue in some way. Some have focused on political ideology. Some have focused on human psychology or the ways in which evolution has led our brain to prioritize short-term rather than long-term threats. Others have called attention to the role of fossil fuel-funded disinformation in undermining public support for action, while others have claimed that climate change is the inevitable consequence of capitalism.

Chien-Yi Lu wants us to step back from the detail and specificity of these claims and confront the larger structure under which these patterns have unfolded and expressed themselves. Drawing most on the last of these suggestions, she focuses our attention on the character of neoliberal capitalism, particularly the development in the last few decades of what she calls SCAMD—states and corporations sitting above market and democracy. Her emphasis is on the way that governments and corporations have worked together, yielding a set of practices, habits, treaties, and laws that place corporate activity above and out of the reach of democracy, thereby making it in effect impossible to enact the changes necessary to stop disruptive climate change.

Her thesis is that it is neither coincidence nor psychological failing, that, despite 30 years of scientific information and warnings, we have failed to address climate change. Rather it is the predictable consequence of neoliberalism, which, as a global ideology, has undermined democracy in ways that have made it impossible for us to address climate change. She does not argue that the advocates of neo-liberalism *intended* to destroy our futures, but she does believe that they intended to undermine democracy, and this has made it impossible for scientists and citizens, and erven those political leaders who wanted to, to enact meaningful responses to climate change. In the introduction to this volume, Chien-Yi Lu writes:

> The original goal of undermining democracy was *merely* to tilt the economic and political systems in such a way that wealth and power would concentrate towards the top. By the time scientists became alarmed by the threat of climate change, however, democracy was crippled enough to keep society from receiving critical information and taking necessary action. (xxi)

Thus, neoliberals, whether intentionally or not, have led us to our present crisis, and neoliberalism stands in the way of its solution. It is not that we have been stupid, but that we have been systematically disempowered.

This is a strong book. Some readers will no doubt take issue with the use of "enemies" as an analytical category. Others might argue with her definition of neoliberalism as "the art—as in 'con artist' of exclusion through inclusion, with upward power and wealth-distribution as its goal." Neoliberalism, like any global movement or ideology, is complex and protean. Any simple definition will no doubt provoke critique, and this particular definition will probably be viewed by many as at worst conspiratorial and at best at least a bit unfair. As Lu herself notes, most neoliberals did not intend to create deadly climate change. What they intended was a set of policies intended to support, expand, and extend markets. At least that is what they would generally claim.

My own view is that at least some of the early neoliberals—Hayek in particular—authentically believed that keeping governments out of markets would increase prosperity while protecting individual liberty. That they paid inadequate attention to matters of equity and were generally dismissive of external costs and market failure does not prove that their intent was other than they expressed. However, it is indisputably the case that, whatever the original intent, the neoliberal regime that has prevailed in the United States and Europe over the past 40 years has had the effect—obvious for some time now—of greatly increasing inequities. It has also had the effect of driving disruptive climate change and other serious environmental and public health harms. It has been obvious for some time that deregulated markets do not and cannot be expected to account for external costs, of which climate change is the most egregious.

A century has passed since the British economist Arthur Pigou pointed out the problem of "incidental uncharged disservices," or what we now call external costs. Pigou suggested that the solution was to tax the offending activity. If the social cost of that activity could be calculated, then one could set the level of the tax to match the level of the social cost. It is from this that we derive the concept of the "social cost of carbon." (Conversely, industries that produced incidental uncharged services, or external benefits, could and should be subsidized. In that sense we could, and should, talk about the "social benefit of education," or "the social benefit of public health.") Since the early 20th century, economists have discussed this issue extensively. They of course argue about the details, yet there appears to be broad agreement among economists that putting a price on carbon—either through direct taxation or emissions trading—could be an important step towards reducing GHG emissions and therefore stopping further climate change. It was this argument that led future U.S. Vice-President Al Gore to propose a B.T.U. (British thermal unit) tax back in the 1980s.

A variation on the theme of carbon pricing is an emissions trading system (ETS). Emissions trading was used successfully in United States in the 1990s to control acid rain in the American mid-west and air pollution in southern California. This success offered a model for how GHG could be controlled and was the basis of the American Clean Energy and Security Act, also known as the Waxman-Markey bill, which was introduced but ultimately failed in the U.S. Congress in the late 2000s.

Carbon pricing, either in the form of emissions trading or taxation, has now been introduced in a number of countries, states, and provinces around the globe. The Canadian province of British Columbia has one of the most well-known systems: its carbon tax, together with a number of other regulatory mechanisms, has been in place since 2008, and it has resulted in a measurable decline in fossil fuel use in that province compared with the rest of Canada.

However, the reduction in British Colombia—about 17%—is not nearly enough. Globally, emissions continue to rise. In many places, the price that has been set on carbon has been so low as to have little effect. As Lu notes, when the European Union (EU) established its ETS, the allowances (or caps) were set too high, so there was little demand for them, and the price fell to almost nothing; the first phase of the system failed. Worse, the ETS gave permits for free to polluters, in effect locking in their right to emissions when what was needed was a system to remove that "right."

The EU continues to work on its system and argues that a planned future decrease in allowances will lead to bigger reductions in the future. But the rate at which those allowances fall is very small: In the period 2013–20 the cap on emissions from power stations and other fixed installations is being reduced by 1.74% every year. For comparison, the IPCC says that we need to cut emissions by 50% by 2030 and phase them out more or less entirely by 2050. Moreover, the EU ETS covers only about 45% of total EU member emissions. In this context, any local "success story" can be viewed as at best a partial measure and at worst giving us a false sense of progress.

Lu insists that we cannot just reassure ourselves that we are making progress, when that progress is insufficient to get us where we need to go. We need to look at the political considerations that led the EU to create a system that was woefully inadequate and the U.S. to do virtually nothing at all. We need to look at the ways in which corporations "sit above democracy," and the ways in which they control markets rather than being subject to them. Readers may not agree with all the answers offered here, but I think that anyone who reads this bold and courageous book will come away agreeing that these are urgent questions.

Dr. Naomi Oreskes
October 2019

Notes

1 https://www.nature.com/articles/s41558-018-0349-9.epdf?shared_access_token=p jzah7DpLvTpHzKC6tJnR9RgN0jAjWel9jnR3ZoTv0P4fU1pUb7k0a5LnOE5Z S6C6a7RZ3NHKtxGsOdgHHB-ipijdyckMqNGCt7v20UfBStXn7urJhqpBnkOr EEg4cBKl28toRgrdYqY_8QwJXcMn236XXSYtNuslXyHqNb8nuU%3D.
2 https://www.youtube.com/watch?v=sqClSPWVnNE.
3 https://sustainabledevelopment.un.org/milestones/humanenvironment.
4 https://public.wmo.int/en/bulletin/history-climate-activities White, R.M., 1979: The World Climate Conference: Report by the Conference Chairman. WMO Bulletin, 28, 3, 177–8.
5 https://www.climate.gov/news-features/understanding-climate/climate-change-at mospheric-carbon-dioxide.
6 https://www.c2es.org/content/short-lived-climate-pollutants/.

Acknowledgment

The habit of listening to podcasts unexpectedly changed the way I see the world. Although it is impossible to list all the programs that have contributed to ideas presented in the book, I do want to mention Democracy Now, Best of the Left, LSE Public Lectures, Alternative Radio, Clearing the Fog, Economic Update with Richard D. Wolff, and The Real News, which I listen to regularly and benefit from tremendously. Independent media such as Truthdig, Truthout, The Intercept, and Common Dreams and organizations such as Corporate Europe Observatory, Corporate Watch, Public Citizen, and Transnational Institute are critical portals for me to see the world behind the one manufactured by the corporate media.

Not all individuals who, through their research, writing, lectures, and/or interviews, have had a profound impact on my understanding of economics, democracy, and neoliberalism are cited, owing to limited space and the structure of the book. I would therefore like to thank Noam Chomsky, Joseph Stiglitz, Ha-Joon Chang, Yanis Varoufakis, Philip Mirowski, Nancy Fraser, Vandana Shiva, Susan George, Wendy Brown, Stedman Jones, David Harvey, and Chris Hedges here for their critical role in subverting my old beliefs about the way the world operates.

Bill McKibben, George Monbiot, Naomi Klein, and of course James Hansen, Kevin Anderson, and Michael Mann are key individuals whose tireless efforts to enlighten the world about climate change eventually reached me and made me understand the importance of taking immediate action. Naomi Oreskes and Erik Conway's work sparked the very first ideas about writing this book.

Animals play a central role in making this book happen. As a strong believer in animal rights, I began to pay close attention to climate change only because it helped to make a strong case against business-as-usual animal consumption. It did not take long before I realized that, in its own right, climate change deserved my full attention. This has not changed the fact that animals remain as the constant and poignant reminder to me of how sickening neoliberalized societies have become. I wish to thank the people who have taught me about and fought side by side with me for the improvement of human–animal relationship. The group of students at the National Chengchi

University in Taipei who fought courageously for the rights of harmless dogs and cats to continue living their lives in the commons of university campus and beyond, encountering countless heartbreaking setbacks in the process, not only occupy a special place in my heart but have also catalyzed the formation of some of the key ideas in this book. Many of the fresh insights of Dr. Abe Lin, a born believer in zoopolis similar to the one envisaged by Donaldson and Kymlicka in *Zoopolis*, have had an important impact on me, making the inconvenient rethinking of the hierarchical human–animal relationship—and their relations with the commons—necessary.

For anyone devoted to academic research, Academia Sinica—the national academy of Taiwan—is heaven. Without the abundant resources that it provides, this book would never have been born. Friends at the Institute of European and American Studies have been a great source of support. Serena Chou has been there for me, both intellectually and emotionally, since the blurred idea about writing this book first emerged. Throughout the process, she has been the one to knock sense into me each time that overwhelming self-doubt made giving up sounded like a viable option.

I am grateful to the Department of the History of Science of Harvard University for hosting me during the summer of 2016 for my research on the book. Special thanks go to Dr. Naomi Oreskes at the Department for her inspiration, guidance, help, and support. My dear friends Fujia Lu and Dan Hu made my stay at Harvard extra fruitful. The discussions that continued deep into the night on topics ranging from this book, the viability of the magic pill I intended to invent, and the old days when we were all still young graduate students remain vividly in my mind and will stay that way for a long time.

Publishing can be an intimidating experience for a first-time book author. Nearly all my worries about Hollywood-portrayed stereotypical editors and publishers were gone after my first contact with Natalja Mortensen at Routledge. She had conveyed nothing but clear guidance, warmth, reassurance, enthusiasm, and encouragement from the very beginning. I would also like to thank Lillian Rand and Charlie Baker for their patience and kind assistance with a very unexperienced author. I am indebted also to the reviewers of my book prospectus for their important comments and suggestions.

Partial funding for research assistance and the editing of the book came from the Ministry of Science and Technology in Taiwan. Jeffrey Cuvilier had been excellent editing and proofreading earlier versions of the book. Pimei Lin and Filip Kszczotek, my wonderful assistants, and Ying-Wen Chen and Wei-Ting Chang, librarians at the Institute of European and American Studies, have been so capable that I am convinced that the words "cannot find" and "need extra time to find" do not exist in their dictionaries.

Very special thanks go to Bernice Maxton-Lee and Graeme Maxton. The kind of support and encouragement that they have provided is unique and irreplaceable. When nearly all my peers remain unfamiliar with or unconvinced of the peril of neoliberalism, their appearance in my life offered me a crucial piece of evidence that I might not, after all, be the one who had gone crazy. Their comments and suggestions in the book are invaluable. Bernice's ability to understand and improve my writing is absolutely magical. The way that her research has influenced mine has already begun to show in this book and will, without any doubt, become more apparent in the future.

Finally, my deepest thanks to my family, for their love, support, and understanding.

List of Abbreviations

AHCI	Arts and Humanities Citation Index
ALEC	American Legislative Exchange Council
APSA	American Political Science Association
BCSD	Business Council for Sustainable Development
BIT	Bilateral Investment Treaty
BRT	U.S. Business Roundtable
CCAP	Center for Clean Air Policy
CCC	carbon-combustion complex
CDM	Clean Development Mechanism
CETA	Comprehensive Economic and Trade Agreement
CPTPP	Comprehensive and Progressive Agreement for Trans-Pacific Partnership
CSR	Corporate Social Responsibility
DG	Directorate General
DiEM 25	Democracy in Europe Movement 2025
EABC	European-American Business Council
ECJ	European Court of Justice
ECT	Energy Charter Treaty
EEC	European Economic Community
EFTA	European Free Trade Association
ELEC	European League for Economic Cooperation
EMS	European Monetary System
EMU	European Monetary Union
ERT	European Roundtable of Industrialists
ETS	Emissions Trading System
FAIRR	Farm Animal Investment Risk & Return
FAO	Farm and Agricultural Organization
FEE	Foundation for Economic Education
FET	fair and equitable treatment
FIELD	The Foundation for International Environmental Law and Development
GCC	Global Climate Coalition
GHG	Greenhouse Gases

GWP	global warming potential
HHS	U.S. Department of Health and Human Services
IARC	International Agency for Research on Cancer
ICS	Investor Court System
ICSID	International Centre for Settlement of Investment Disputes
IEA	Institute of Economic Affairs *and* International Energy Agency
IHS	Institute for Humane Studies
IMDb	Internet Movie Database
IMSC	International Market Support Committee
IPPC	International Pollution Prevention and Control
IPCC	Intergovernmental Panel on Climate Change
ISDS	Investor-State Dispute Settlement
ISI	Institute for Scientific Information
LSE	London School of Economics
MIC	Multilateral Investment Court
MPS	Mont Pelerin Society
NAFTA	North America Free Trade Agreement
NRDC	Natural Resources Defense Council
NYU	New York University
OECD	Organization for Economic Cooperation and Development
OPEC	Organization of the Petroleum Exporting Countries
SCAMD	States and Corporations sitting Above Market and Democracy
SCI	Science Citation Index
SEA	Single European Act
SGP	Stability and Growth Pact
SSCI	Social Sciences Citation Index
TABC	Trans-Atlantic Business Council
TABD	Trans-Atlantic Business Dialogue
TENs	Trans-European Networks
TNCs	Transnational Corporations
TTIP	Transatlantic Trade and Investment Partnership
UNCITRAL	United Nations Commission on International Trade Law
UNFCCC	United Nations Framework Convention on Climate Change
UNICE	Union of Industrial and Employers' Confederations for Europe
USDA	U.S. Department of Agriculture
USMCA	U.S.-Mexico-Canada Agreement
WHO	World Health Organization
WMO	World Meteorological Organization
WVF	William Volker Fund

Introduction

Chien-Yi Lu

A key component of climate action involves identifying, understanding, exposing, and weakening its enemies—those who have a vested interest in denying climate change and downplaying its threat. At every opportunity, these enemies smear, stall, block, neutralize, divert, crowd out, and sabotage honest and effective mitigation policies. The most easily identified enemy of climate action is the fossil fuel industry. Politicians and financial institutions benefiting from the protection of the fossil fuel industry are also well-known enemies of the planet. Oreskes and Conway use the term "carbon-combustion complex" (CCC) to denote:

> a network of powerful industries comprising fossil fuel producers, industries that served energy companies (such as drilling and oil field service companies and large construction firms), manufacturers whose products relied on inexpensive energy (especially automobiles and aviation, but also aluminum and other forms of smelting and mineral processing), financial institutions that serviced their capital demands, and advertising, public relations, and marketing firms who promoted their products. (2014: 36–7)

Such a complex, however, was not able simply to form into a powerful alliance one day and become capable of defending the destabilization of the climate on the next. Some preconditions must exist for this alliance to dominate other forces in society. Among such preconditions include the malfunctioning of democracy, because it is incompatible for the CCC to thrive alongside a well-functioning democracy.

To add to our understanding of the enemies of honest and effective mitigation measures, this book investigates the organized efforts which, by undercutting, bypassing, and overriding democracy, have a direct impact on climate change. If the organized efforts undermining democracy worked as a larger force behind the CCC, safeguarding a favorable environment for anti-mitigation operations while shielding it from public scrutiny, then these organized efforts must be treated as the fundamental enemy of climate action. The organization of such efforts started long before scientists noticed that the planet was warming. The original goal of undermining democracy was *merely* to tilt the economic and political systems in such a way that wealth and power would concentrate towards the top. By the time that scientists became alarmed by the threat of climate change, however, democracy was crippled enough to keep society from receiving critical information and taking necessary action. In this sense, the stable climate is just one collateral damage of these organized efforts among many, including the environment at large and the health of the public.

The organized efforts of undercutting, bypassing, and overriding democracy owed much of their success to the magic tool of neoliberalism. Neoliberalism has many definitions. The most common is the marketization and commodification of everything under capitalism. In this book, I define neoliberalism as *the art—*as in "con artist"*—of exclusion through inclusion, with upward power- and wealth-redistribution as its goal.* The discrepancy between what neoliberals claim to believe, e.g., "we are all in it together," and what they *actually* believe, i.e., the dispensability of large portions of society, makes it dangerous to treat neoliberalism as a genuine economic theory, school of thought, or ideology. Immeasurable amounts of time, energy, and talent have already been wasted on engaging sincere (on the part of non-neoliberals, that is) debate with neoliberals *as if* they were *honest* theorists, thinkers, scholars, think tank experts, or statesmen, when actually, the core feature of neoliberalism is deceit. In fact, this phenomenon pleases neoliberals very much. By elevating neoliberalism to such an undeserved rarefied level, these debates obscure the true nature of neoliberalism and help to prolong its incumbency as the dominant order.[7] A simple principle of the art of exclusion through inclusion was to emphasize individual freedom of choice at the tangible and concrete level—be it McDonald's or KFC; iPhone or Android. Individual choices at this level, however, come at the very serious price of losing meaningful participation at the abstract and higher level, where decisions regarding distributional rules are concerned. The false perception of being included has lulled the global democratic citizenry and kept it from detecting the hollowing-out of democracy. Zombie democracy, functioning as lip-service, catch-phrases, and rubber stamps, is the best friend of neoliberalism, reinforcing its legitimacy. It is hardly surprising, then, that alongside planetary warming has come social and economic misery at a mind-boggling level in the world's most advanced democracies, for reasons that are independent of climate disruption. According to the World Bank, in 2015 some 3.7 million Americans and 100,000 British people lived in extreme poverty, defined as living on less than US$1.90 a day (The World Bank, 2018: 45). In 2016 about 40.6 million (or 12.7% of the total population) Americans and 13.9 million (22%) British people lived in poverty (U.S. Census Bureau, 2017: 12; Joseph Rowntree Foundation, 2017: 10).

The art of exclusion through inclusion has its roots in modern economics. The birth of modern economics dates back to at least the turn of the 18th and 19th centuries, when economists such as David Ricardo and Thomas Malthus began formulating economic theories to justify the exclusive distribution of finite resources. The clash between these exclusionary dynamics and its counter-movement—the inevitable consequence of the social pain caused by exclusion—culminated in two World Wars in the 20th century. The catastrophic experiences of these wars led some commentators (most notably, in the context of this study, Karl Polanyi) to believe that societies had learned their lesson and that aggression in the guise of economic theory no longer had a place in the modern world. Such optimism, however, turned out to be misguided. The renewed alliance between those eager to exclude and those with the talent to package such exclusion in polite economic niceties began to form as soon as Europe emerged from the Second World War. The unholy alliance between the thinking of Austrian economist Friedrich Hayek and his sympathetic business funders plays a decisive role in the climate challenge that we now face. With this alliance as a starting point, the organized efforts for undermining democracy have snowballed, perpetuating and entrenching the power asymmetry between the very rich and the rest of society. Today, it is precisely this deeply entrenched power asymmetry that is at the core of the continued stalling of effective climate mitigation. The business-funded victory of Hayek's art of exclusion over Keynes' economic theories, which were inclusive in nature and thus incompatible with neoliberal ambitions, is therefore one of the most important and unfortunate developments in human history. Far from being a "free competition of ideas," the great "debate" between Hayek and Keynes needs to be seen in the same light as the "debate" between climate deniers and genuine climate scientists.

Business-funded economic theories aimed at justifying exclusion would soon have run into obstacles from neighboring social science disciplines, if a "corrective" mechanism had not been erected. Political science—the discipline that specializes in the study of democracy, for instance—might have kept at bay the erosion of democracy by neoliberal economic policies. To eliminate this possibility, business funders and neoliberal economists went to work reshaping and redefining disciplines, from law to political science. The scant attention that political scientists have directed toward climate change, despite it being essentially a political problem, can easily be explained by the neoliberalization of the discipline. Far from being just passively indifferent to the climate emergency, political science, having been neoliberalized, has been an active *promoter* of the neoliberal way of thinking and thus an active *contributor* to impediments of mitigation.

As business funders and neoliberal social scientists were busy constructing and selling theories that aimed at steering the dominant discourse clear of actual reality, John Kenneth Galbraith, in *The New Industrial State*, recorded the truth and explained the way that this discourse worked to hide truth. Based on his analysis and insights, this book uses the acronym, SCAMD—States and Corporations sitting Above Market and Democracy—to denote the global neoliberal structure. In the SCAMD structure, the state and large corporations are often merged into one actor, wielding power not from *within* the market and democracy but from *above* them, overlooking the rest of society. The existence of the SCAMD structure explains the remarkably effective disembedding of markets from society.

A quintessential example of the state being interwoven with corporations in masterminding an artificial market which facilitated and extended the upward redistribution of power and money from ordinary citizens to SCAMD elites is the cap and trade system of the European Union (EU), known by its acronym, ETS (emissions trading system). Although the EU has long been portrayed as the poster child of climate mitigation, the fascinating story of SCAMD elites sabotaging an EU-wide carbon tax and putting in its place the lucrative ETS reveals the neoliberal—and hence mitigation-impeding—nature of the EU. As Perry Anderson notes, "the self-satisfaction of Europe's elites, and their publicists," is such that the Union is "widely presented as a paragon for the rest of the world, even as it becomes steadily less capable of winning the confidence of its citizens, and more and more openly flouts the popular will." To curb the degradation of democracy, it is imperative to abandon the illusion that "within the Atlantic ecumene Europe embodies a higher set of values than the United States, and plays a more inspiring role in the world." The construction of the EU, notwithstanding its peace-promoting founding principles, was a crucial step in building the global SCAMD structure. Many of the key players in the Transatlantic elite network who had crucial roles in discrediting Keynes' inclusive economic theories were the very figures who initiated and designed the structures and institutions of the EU. This explains the striking similarities between the Maastricht Treaty and the Chilean constitution, which SCAMD elites helped to draft. Instead of being the poster child of climate mitigation, the EU is in fact taking the lead, through deliberately structured investment and trade agreements, in outlawing government regulations (which demonstrably are by far the most effective mitigation tool) that displease multinational corporations through investment and trade agreements. This "Investor–State Dispute Settlement" (ISDS) mechanism is inserted not only in bilateral investment agreements, but also in multilateral treaties such as the Energy Charter Treaty, which the EU is actively seeking to expand.

Given the enormous power and manipulative capability of the SCAMD elites, climate actions that fail to address the fundamental problem, namely, the existence of the SCAMD structure, are bound to fail. Initiatives and movements such as the Green New Deal and the Extinction Rebellion are crucial for the survival of the planet precisely because implicit in their action plans is the undermining of the anti-democratic SCAMD structure. Beyond such actions in the political and civic realms, groups such as the Fossil Fuel Divestment Movement are critical flanking forces, owing to their ability to engage with individuals whom political and civic initiatives are less capable of reaching. Countering the force of the SCAMD structure requires strategies to address the way in which the state and large corporations merge, the central feature of the structure and a key cause of the abject state of democracy today. In the same way that SCAMD elites leveraged the fluid relationship between the market and the state, nurtured by corporate–government alliances, SCAMD-countering strategies such as Fossil Fuel Divestment merge citizen power with consumer and investor powers. While recognizing the importance of the market, the movement is far from being an endorsement of neoliberalism. Markets are perfectly serviceable human institutions when embedded in society. This kind of market—the only kind that is sustainable, according to Polanyi—can be made real; and rebellion aligning citizen, consumer, and investor positions is a step in that direction. Rather than engaging in blind market-bashing and voluntarily forsaking the battleground of the market in the struggle for survival, therefore, climate activists should double down on catalyzing change through synchronized citizen-consumer-investor attacks on the SCAMD structure.

Overwhelming scientific evidence on the astonishing contribution of the livestock industry on climate change makes expediting the bursting of the meat bubble a perfect candidate for doubling down on synchronized citizen-consumer-investor attacks on the SCAMD structure. Diet adjustment is a powerful tool for tackling climate change because neither public infrastructure nor mass mobilization is required for individuals to switch to a climate-friendly diet. A utilitarian adjustment calculated to toe the line of animal exploitation at the "1.5°C consistent" fine line, however, is merely a technical fix rather than a systemic overhaul of neoliberalism-compelled climate emergency. It is as reasonable as robotically calculating the amount of fossil fuels having been burned and that can still be burned before destroying the planet, without investigating the fundamental question of why and how the world came to depend so heavily on fossil fuels, which nurtured the situation and which continue to block transition. Like the human–climate relationship, the human–animal relationship today is shaped by organized efforts aimed at undermining, bypassing, and superseding democracy. It is not an isolated issue separated from climate change, but rather an integral part of the neoliberalism-compelled phenomenon. A core message of the book is that climate change is a problem that has emanated *not* from burning fossil fuels, but from the success of the neoliberal art of exclusion. As a corollary, mitigation strategies that tackle not the *symptoms* but the *root cause* of climate change involve a wholesale and fundamental rejection of neoliberalism, including its exclusionary and utilitarian way of treating not only people of different economic status, race, and nationality, but also nature, the climate, and non-human species. Such a new perspective makes it instantly clear that the role of non-human sentient beings is not to satisfy neoliberalism-enhanced human cravings any more than the function of the 99% is to serve the insatiable desire of the 1%. Mass-scale unspeakable animal suffering is a form of collateral damage of the neoliberal exclusionary scheme, just as the disrupted climate is. The neoliberal business-as-usual treatment of animals is profoundly contradictory to root-cause-tackling of mitigation strategies. Neoliberalism has been extraordinarily successful in stigmatizing the animal-respecting vegan diet. It is high time that the vegan diet is de-stigmatized while efforts are made to stigmatize carnism instead.[2]

The book is organized as follows: Chapter One examines the relationship between climate change and democracy. The academic work on this relationship has swung like a pendulum between seeing democracy as an obstacle to environmental conservation and viewing more democracy, rather than less, as the key to survival. After reviewing the literature, which seems to conclude on a pessimistic note regarding the compatibility of democracy with mitigation, I point out that the problem is not democracy *per se*, but *the state* of democracy, particularly in advanced democratic countries. The decadence of democracy, rather than being the sum of mindlessness and sloppiness scattered here and there, was in fact the intended consequence of organized efforts that have painstakingly plotted to undercut, bypass, and override democracy.

Chapter Two takes the discrediting of Keynes' inclusive economic theories, which treat all members of society as indispensable, as the starting point of the organization of efforts aimed at undercutting, bypassing, and overriding democracy. It highlights the critical role that money played in the "free" competition of ideas between Keynes and Hayek. The very reason that business funders lined up behind Hayek's theories and ensured his eventual victory over Keynesian economic theories in the policy world was the exclusionary nature of Hayekian economic thought. It was precisely to hide this exclusionary nature that the remarkably sophisticated tool of neoliberalism, which sugar-coated exclusionary policies in self-regulating market discourse, was invented with business money. Neoliberalism tricked the public into endorsing a socioeconomic order arranged according to an exclusionary logic, guaranteeing a widening power asymmetry between the wealthy and influential and the rest of society, just as its designers intended. This power asymmetry goes a long way to explain many of the intractable problems of our time, most devastatingly, climate change. As a result, it is as dangerous—if not more—to treat neoliberalism as a legitimate and honest economic theory as it is to treat climate denial as true science.

Chapter Three delves into the conquering of political science by neoliberal economists. The indifference of political science as a whole to the climate emergency can easily be explained by the neoliberalization of the discipline. Beyond this passivity, by unquestioningly embracing the neoliberal "methodological individualism," political science has played an active role in conforming the world to the neoliberal ideal. The Chapter examines in detail the critical role played by neoliberal economist James Buchanan in revolutionizing political science in order to stop the study of democracy from impeding the neoliberal project to undercut democracy. In time, political science went through a "scientific revolution" and embraced the "rational" (utilitarian, calculating, self-interest maximization) approach that always took the individual, as opposed to the community, society, or the commons, as the starting point of inquiry. Like a toothpaste advertisement, the claim of being scientific in academia has the effect of donning a lab coat, enhancing the authority and credibility of the scholar conveying his view. As Jon Elster alluded to, public choice or rational choice theories were above all *normative* theories. They work as dog whistles, shaping human behavior into conformity with neoliberal ideals, as the inventors of such theories had intended.

The reality behind the reality constructed by neoliberals is the topic of Chapter Four. I rely on John Kenneth Galbraith's *The New Industrial State* for the analysis of the extraordinary state–corporate symbiotic relationship that blossomed in the post-war period in the U.S. As a high-level civil servant in the U.S. government, Galbraith had the chance to see first-hand how the market really worked and concluded that it was not self-regulating. Quite to the contrary: by the 1960s the accurate name for the economic system operative in the most advanced free-market society would be the "planning system." Galbraith explained that the market, rather than being self-regulating, had become subordinated to large corporations which, with the help of the state, rose as the key allocator of values and resources. This development necessitated the active nurturing of the mainstream discourse aimed at keeping public perception of reality far distant from actual reality. In the nurtured discourse, the market was in full control, large corporations were subservient to the market just as small and medium businesses were, and the state was completely separate from large corporations in its goal and identity. This discourse sustained the myth that the sequence in market activities began with what sovereign consumers demanded, when in reality the sequence started with what producers, with the help of the state artificially removing a significant amount of market uncertainties, wanted to supply.

Although Galbraith's discussion of the "planning system" concerned the U.S. alone, large corporations' subordination of the market knows no boundaries. Chapter Five focuses on the poster child of climate mitigation, the EU, and analyzes its climate strategies by situating them within the context of an expanding neoliberal planning system. The first half of the chapter deals with the nature of the EU: If the self-regulating–market discourse was essentially a neoliberal tool aimed at undermining democracy and facilitating upward wealth and power redistribution, then the establishment of the EU, which was founded on the self-regulating-market logic, might be seen as serving the purpose of facilitating upward redistribution. Such an understanding helps to make sense of the development of the EU's climate policies, which is the focus of the second half of the chapter. It investigates why a Union-wide carbon and energy tax failed to be adopted as a EU policy while the corporate-serving cap and trade system, the emissions trading system (ETS), became—and firmly remains—the backbone of the EU's climate policy.

Even though the EU-wide carbon tax did not survive corporate sabotage, and the EU ETS has been proven to be counter-productive for climate mitigation, the EU and its member states still have the alternative of government regulation as an effective tool for climate mitigation. Chapter Six explains how the option of regulation is increasingly being outlawed by the signing of trade and investment treaties. The EU is among the most zealous in pursuing and imposing such treaties, which are essentially designed to help investors to fend off government regulations, including mitigating measures, that run counter to their business interests. Through the instrument of ISDS (Investor-State Dispute Settlement) embedded in treaties such as Energy Charter Treaty, the Comprehensive Economic and Trade Agreement and the Transatlantic Trade and Investment Partnership, the SCAMD elites are completing their task of subordinating markets and democracy alike.

Chapters Two through Six provide a historical description of the neoliberal art of exclusion through inclusion. They establish that the basis of the current global social economic system is a massive public relations scheme aimed at disguising the exclusionary grabbing impulse of the few at the top. Chapter Seven looks forward and reflects on strategies for pushing back the neoliberalization of democracy. The importance of civic action in fighting back neoliberalism is too obvious to merit discussion. It is the critical question of the place of the market that warrants consideration. The chapter explicates the crucial role that the market can play in de-neoliberalizing democracy and curbing climate change. Apart from the fact that time-constrained climate mitigation does not have the luxury of abandoning the market as a key battleground for the struggle, particularly given how marketized all aspects of human lives have already become, it is crucial to recognize that the market is not something that must be *done away with* in the fight against neoliberalism. Rather, the urgent task in the fight involves *re-embedding* the market in society. Citing overwhelming scientific evidence, Chapter Seven identifies the expedition of the bursting of both the carbon and the meat bubbles as the most urgent market re-embedding missions in saving the planet. The discussion of the meat bubble brings the book full circle to the question of democracy, neoliberalism, inclusion, and exclusion. Highlighting the role of neoliberalism in redefining the human–animal relationship in a way that destroys the environment, disrupts the climate, consolidates increasingly asymmetrical power structure, and justifies violence, exploitation, and exclusion, the Chapter reaffirms the message that climate change is merely one component of a larger organic crisis with its origins in neoliberalism. Instead of addressing symptoms one by one, issue by issue, surviving the organic crisis calls for a wholesale rejection of neoliberalism and for rebuilding an inclusive society where none, even non-human sentient beings, is reduced to servitude.

Notes

1 It is important to point out that those who support the current economic order but are unaware of the hidden exclusionary agenda of neoliberalism are *not* neoliberals as defined by this book. A neoliberal readily accepts that some

members of the society are dispensable and knows to always conceal this belief unless in conversation with other neoliberals. Here, Hay and Rosamond's differentiation between the internalization of a discourse on globalization and the "intentional, reflexive and strategic *choice* of such a discourse" is relevant (Hay and Rosamond, 2002: 150). At the same time, neoliberalism was not developed single-handedly by a closed conspiratorial circle. Although the term "neoliberalism" refers to a complex nexus of ideas, tactics, institutions, and individuals, not all such individuals particulate in each and every activity discussed in the book, even though all are implicated in at least some of the activities. I am indebted to Naomi Oreskes for reminding me to add this clarification.

2 Melanie Joy coined the term "carnism," which she defines as "the belief system that conditions us to eat certain animals." Under carnism, "choices appear not to be choices at all," because it is "a particular type of ideology… especially resistant to scrutiny" (2010: 30).

References

Hay, Colin and Ben Rosamond. (2002). "Globalization, European Integration and the Discursive Construction of Economic Imperatives." *Journal of European Public Policy.* 9:2, 147–167.

Joseph Rowntree Foundation. (2017). *UK Poverty 2017—A Comprehensive Analysis of Poverty Trends and Figures.* York: Joseph Rowntree Foundation.

Joy, Melanie. (2010). *Why We Love Dogs, Eat Pigs, and Wear Cows—An Introduction to Carnism.* San Francisco: Conari Press.

Oreskes, Naomi and Erik Conway. (2014). *The Collapse of Western Civilization—A View from the Future.* New York: Columbia University Press.

Orwell, George. (1949). [1996]. *Nineteen Eighty-Four.* London: Secker & Warburg.

U.S. Census Bureau. (2017). *Income and Poverty in the United States: 2016.* Washington, DC: U.S. Census Bureau.

World Bank. (2018). *Poverty and Shared Prosperity 2018—Piecing Together the Poverty Puzzle.* Washington, DC: World Bank.

1 Is Democracy in the Way?

The fact that the scientific knowledge on the human contribution to climate change entered human society through the most advanced democratic societies should have been a cause for celebration. Given the congruence of climate mitigation and public interests, the problem of climate change should have been considered solved decades ago. Several decades of inaction later, however, arguments are proliferating that democracy is exactly the reason for inaction.

In *The Collapse of Western Civilization*, historians Naomi Oreskes and Erik Conway travel to the future to look back and offer a forensic analysis on the climate-induced Great Collapse of Western Civilization of 2074 (2014: 63). The future historians' forensic report states that "[a]s the devastating effects of the Great Collapse began to appear, the nation-states with democratic governments… were at first unwilling and then unable" to deal with the crisis. These democratic governments realized that they had no "infrastructure and organizational ability to quarantine and relocate people" as "food shortages and disease outbreaks spread and sea level[s] rose." In China, where there was centralized government, the crisis was handled much more adequately, leading to survival rates exceeding 80%, a development that "vindicated the necessity of centralized government" (2014: 51–2). The gist of *The Collapse of Western Civilization* is not about critiquing democracy *per se* but a warning against the stubborn inaction mandated by market fundamentalism that has hijacked Western democracies.[1] In their previous book, *Merchants of Doubt*, Oreskes and Conway documented the way that climate deniers sowed the seeds of doubt about climate change and successfully staved off implementations of mitigation measures. For the authors, the anticommunist ideology that had kept actors vigilant about government encroachment in the marketplace occupied a central place in climate denial (2014: 69). Ironically, this sort of ideology-informed calculation meant that preventative action was blocked, increasing the risk that disruptive climate disasters would eventually necessitate the suspension of democracy and legitimating the sort of heavy-handed authoritarian interventions that the conservatives most abhorred (2014: 52; 69).

An appeal to suspend democracy for the sake of survival can be found in *The Climate Change Challenge and the Failure of Democracy*, where

Shearman and Smith argue that liberal democracy is incompatible with the urgent necessity to prevent catastrophic climate change. The vested interests of politicians, corporations, and media lie in continuing with business as usual and in keeping the public ignorant. Instead of bottom-up reforms to improve democracy and bring about sensible climate policies, Shearman and Smith see a transformation into authoritarian regimes as the only responsible way forward when faced with the extreme ecological stress of climate change. They point out that, as Plato foresaw, those in power in a democracy are seldom able to resist the demands of the populace for long, but as a mass, the populace is seldom able to focus on complex problems and to perceive threats that lie over the horizon. Hence, those able to see further—scientists, experts, and the knowledgeable—should be entrusted with steering the course while there is still time to avoid disaster. It is only under a benign authoritarian rule of the knowledgeable that a saner, fairer, and more rational means of weighing social goods against evils can be introduced (Shearman and Smith, 2007).

The doubt about the ability of democracy to handle climate challenges is palpable from the intellectual Left as well. Eric Hobsbawm offered a three-fold explanation for his pessimism. To begin with, many of the strategies needed to avoid climate change would be extremely unpopular and therefore difficult to implement in a democracy. As a result, even as "the impact of human action on nature and the globe has become a force of geological proportions," "no support will be found by counting votes" for measures required for mitigating these problems. Moreover, given that nature is border-blind, even if voters of some democratic states were sensible, the political mechanisms available to human kind in the 21th century are "effectively confined within the borders of nation-states" and "dramatically ill-suited" to deal with problems lying beyond their range of operation (2007: 113). Finally, democratic national governments are not the only relevant organizational entities that can have an effect on an increasingly globalized and transnational world. "A growing part of human life now occurs beyond the influence of voters, in transnational public and private entities that have no electorates, or at least no democratic ones." Thus, "[d]emocracy, however desirable, is not an effective device for solving global or transnational problems" (2007: 118).

This wave of academic literature that questions the compatibility of democracy with timely and effective climate mitigation resonates with works dating back to the 1970s that focused on the role of democracy in environmental conservation. In *An Inquiry into the Human Prospect*, Heilbroner set to answer, in a world plagued by problems such as rapid environmental degradation, "is there hope for man?" Writing in 1974, he highlighted that:

> the amount of CO_2 in the air is expected to double by the year 2020… sufficient to raise surface temperatures on earth by some 1.5° to 3.0°… bring[ing] sea levels above the level of the land in the populous delta areas

of Asia, the coastal areas of Europe, and much of Florida. Long before that it is feared that the rise in temperature would have irreversibly altered rainfall patterns, with grave potential effects. (1980 [1974]: 72)

With the approaching of the depletion of natural resources, Heilbroner expressed deep doubt about the ability of the democratic form of government in ensuring the survival of mankind.

[C]andor compels me to suggest that the passage through the gantlet ahead may be possible only under governments capable of rallying obedience far more effectively than would be possible in a democratic setting. If the issue for mankind is survival, such governments may be unavoidable, even necessary. (1980: 130)

This pessimism stems from the unavoidable transition of capitalism from its expanding form to a stationary one under severe scarcity of resources, as *"whether we are unable to sustain growth or unable to tolerate it…*, it seems beyond dispute that the present orientation of society must change" (1980: 110, original emphasis). Social tensions will inevitably rise when scarcity-propelled stationary or even slow-growing capitalism renders infeasible the usual method of appeasing the lower and middle classes by further deepening the grab into the nature to improve their economic positions, leaving the diminishing of the incomes of the upper echelons of society the only option (1980: 102). Given the widespread belief that "centralized authority will cope with crisis and unrest more 'successfully' than less authoritarian structures" and the historic pattern in democracies where "the pressure of political movement in times of war, civil commotion, or general anxiety pushes *in the direction of authority*, not away from it," (1980: 128–9, original emphasis) Heilbroner concluded that intolerable socioeconomic strains will eventually exceed the capabilities of representative democracy, leading governments of these societies to resort to authoritarian measures (1980: 106).

Similarly, Ophuls contended that under conditions of ecological scarcity, if individuals are allowed to pursue their self-interest "unrestrained by a common authority," the result is bound to be "common environmental ruin" (1977: 151). Accordingly:

the individualistic basis of society, the concept of inalienable rights, the purely self-defined pursuit of happiness, liberty as maximum freedom of action, and laissez faire itself all become problematic, requiring major modification or perhaps even abandonment if we wish to avert inexorable environmental degradation and eventual extinction as a civilization. (1977: 152)

To him, the only solution is "a sufficient measure of coercion;" and "democracy as we know it cannot conceivably survive" (1977: 151–2).

In the same vein, Ophuls and Boyan (1992) talked about the crucial role that "ecological mandarins" must play under resource scarcity. Concurring with Robert Dahl's point that "a reasonable man will want the most competent people to have authority over the matters on which they are most competent" (Dahl, 1970: 58), Ophuls and Boyan emphasized that "under certain circumstances democracy *must* give way to elite rule," and "the more closely one's situation resembles a perilous sea voyage, the stronger the rationale for placing power and authority in the hands of the few who know how to run the ship" (Ophuls and Boyan, 1992: 209, original emphasis). Given that ecology is esoteric and that only those with talents and training are qualified as specialists, "a class of ecological mandarins who possess the esoteric knowledge" is required to run the "ecologically complex steady-state society" well. Such a society

> will not only be ostensibly more authoritarian and less democratic than the industrial societies of today (the necessity of coping with the tragedy of the commons would alone ensure that), but it may also be more oligarchic as well, with full participation in the political process restricted to those who possess the ecological and other competencies necessary to make prudent decisions. (1992: 215)

The portrait of democracy—or the selfish and ignorant general public—as the obstacle to climate mitigation or environmental preservation, however, does not necessarily square with the reality. As Oreskes and Conway point out, public opinion polls in 2007—the year when the IPCC declared anthropogenic warming to be unequivocal—showed that "a majority of people—even in the recalcitrant United States—believed that action was warranted" (Oreskes and Conway, 2014: 7). Likewise, Krosnick and MacInnis find it inappropriate to attribute a lack of legislation on climate mitigation in the U.S. to a lack of public support. Contrary to the claim that it is the selfish and ignorant general public that is blocking necessary legislation, Krosnick's and MacInnis's study demonstrates that a large majority of Americans have supported a variety of policies aimed at reducing greenhouse gas (GHG) emissions, and the policy endorsement has been consistent across years and across scopes and types of policies (2013: 26). In June 2010 a total of 76% of Americans surveyed answered positively to the question whether the U.S. government should limit U.S.-business-generated GHG emissions. In late 2010 and in 2012 that figure was 74% and 77%, respectively. In 2006, as many as 86% of the respondents said that the government should act to reduce utilities' emissions. This figure increased to 87% in 2007 and stayed in the 70%–80% range even after the financial crisis (2013: 28). Surveys done in the months leading up to the 2008 presidential election showed that 59% of Americans supported a cap and trade system. When given the information that a similar system had worked in other cases, respondents endorsing the system rose to 74% (2013: 30). In sum,

Krosnick and MacInnis found that public support for legislative progress in GHG reduction "seems to be not only present but prevalent." To explain why legislative action did not reflect this strong public preference, they point to the possibility that "legislators have thus far chosen to ignore the will of their constituents when voting on legislation in this arena" (2013: 38). Similarly, in Europe, survey data in the early 1990s already demonstrated that "the weight of public opinion [was] indeed tilted firmly in the 'green' direction" (Witherspoon, 1996: 43).

The propaganda that the public is too stupid and selfish to make enlightened decisions about the environment is not the only misguided notion that defies empirical evidence. Ideas about an authoritarian alternative to democracy are characterized by ill-justified confidence that an authoritarian environmental regime should be benign. As Passmore pointed out, "[t]he view that ecological problems are more likely to be solved in an authoritarian than in.... a liberal democratic society rests on the implausible assumption that the authoritarian state would be ruled by ecologist-kings" (Passmore, 1974: 183). Not only are authoritarian leaders unlikely to be "green philosopher kings," but policies derived from such leaders are also unlikely to be effective. While expertise is no doubt relevant to environmental decision-making, it is not sufficient. Effective decision-making must involve both expertise and the views of those who are most affected (Paehlke, 1988: 296; 1996: 18). Citing Dewey, Taylor argued "[i]f policy is defined and controlled solely by experts, elites, ideological minorities or philosopher kings, it necessarily represents the interests, concerns and values of only a fraction of the community." As a result, there seems no alternative to subjecting environmental policies to debate and deliberation of democratic citizens, warranting an attempt by the political community "not only to solve particular problems, but to define its priorities, concerns, values, its very moral character" (1996: 101).

These strong reservations about authoritarian environmental regimes echo what Caldwell observed already in 1963:

> Scientists may one day tell us what kinds of environment are best for our physical and mental health, but it seems doubtful if scientists alone will be able to determine the environmental conditions that people will seek. There will surely remain an element of personal judgment that cannot be relegated to the computer. (1963: 139)

In view of these irreplaceable features of democracy, Paehlke concluded that authors such as Heilbroner and Ophuls had underestimated democracy while overestimating authoritarianism, and that the answer to future environmental and resource problems might be found in *more* rather than *less* democracy. "Democracy, participation, and open administration carry not only a danger of division and conflict, but as well perhaps the best means of mobilizing educated and prosperous populations in difficult times" (Paehlke, 1988: 294–5).

Thanks to the works of Paehlke and others like him, Dryzek declared in 1996, "[i]f two or more decades of political ecology yield any single conclusion, it is surely that authoritarian and centralized means for the resolution of ecological problems have been discredited rather decisively" (1996: 108).[2] Referring to theorists contemplating or predicting authoritarian and centralized ecological solutions in dealing with the clash between capitalism and ecology as a "theorist[s] of ecological apocalypse" (1996: 114), Dryzek dismissed such an approach as a *"strategy* of awaiting the apocalypse." Rather than waiting for disasters to happen, society needed to adopt a pro-active approach that pursues ecological values in democratic fashion (1996: 115, original emphasis).[3]

Several decades after the alleged discrediting of the centralized ecological solutions and Dryzek's rejection of a "strategy of awaiting the apocalypse," the world is now no longer *awaiting* the apocalypse but at the initial stage of *experiencing* it. The new *status quo* unsurprisingly sets off a renewed round of assessments on authoritarian climate solutions, as was discussed in the first part of this chapter. Like the previous wave of debate, voices cautioning against romanticizing benevolent despotism surfaced to highlight, as does Nico Stehr, that "knowledge of nature must always enter society through politics (whether democratic or authoritarian)—through decisions about... who gets what, when, how" (Stehr, 2016: 42), echoing Caldwell's point made half a century earlier.

What lessons can we learn from these recurring debates concerning democracy and conservation? Is democracy really in the way? Or is the opposite view calling for more rather than less democracy the key to survival? As numerous authors mentioned in this text have alluded to, these are in fact empirical questions, the answers to which are contingent on the *kind* of democracy being practiced on the ground. Here, "the kind of democracy" refers not to the categorization of representative vs. deliberative or parliamentarian vs. presidential democracies. As useful as typologies like these are for understanding varied policy outputs from varied constitutional designs, such categorization, by focusing on the institutional framework of democratic systems, does not capture the key feature of democracy today.

Whether representative or deliberative, and whether parliamentarian or presidential, democracy is a political arrangement where the people are the ones that govern the society. It is different from an oligarchy, where a small group of people governs, and plutocracy, where the wealthy have more say than the poor. What democracy is supposed to be able to achieve is to keep the system from sacrificing the wellbeing of many to serve the interests of the few. Today, the most advanced democratic systems are still to be found in the West, including the United States and the EU. Delegations from these well-established democracies are routinely dispatched to observe elections in newly democratized countries, offering opinions on how these young democracies should go about consolidating their democracies. It seems unproblematic, then, to assume that "democracy" is how one would label these Western countries despite all the criticisms and challenges that they face.

According to a study by Gilens and Page on American democracy, economic elites and organized groups representing business interests have substantial, independent impacts on U.S. government policy, which stands in sharp contrast to the little or no independent influence exerted by average citizens and mass-based interest groups. When the preferences of economic elites and organized interest groups are controlled for, the preferences of the average American citizen appear to have "a minuscule, near-zero, statistically non-significant impact upon public policy." In short, when a majority of citizens disagree with economic elites and organized interests, these citizens generally lose (Gilens and Page, 2014: 576). The authors thus conclude that while Americans enjoy many features of democratic governance, including regular elections, freedom of speech and association, and a widespread (even though still contested) franchise, "America's claims to being a democratic society are seriously threatened," given that "policymaking is dominated by powerful business organizations and a small number of affluent Americans" (Gilens and Page, 2014: 577). To many, it is apparent that the American system is "increasingly coming to resemble a plutocracy" (Milanovic, 2016: 199).

That democracy on the one hand and the extreme concentration of wealth on the other cannot comfortably co-exist is a simple truth, which former U.S. Supreme Court Justice Louis D. Brandeis repeatedly emphasized in the early 19th century (Campbell, 2013). A century later, Thomas Piketty used empirical data to demonstrate that, in recent decades, the concentration of capital had attained a level so high that it was "potentially incompatible with the meritocratic values and principles of social justice fundamental to modern democratic societies" (2014: 26). In *Ruling the Void*, Peter Mair documented the process through which Western democracy was hollowed out. Referring to E.E. Schattschneider's *The Semi-Sovereign People* (1960), which examined the over-representation of affluent interests in pluralist societies, Mair argued that even semi-sovereignty was slipping away today; the people were becoming *non*-sovereign instead. Focusing on contemporary Europe and using data on voter turnout, electoral volatility, and party membership, Mair found that "[w]hat we now see emerging is a notion of democracy that is being steadily stripped of its popular component—easing away from the demos" (2013: 2). During the global financial crisis, it became more difficult to conceal elite determination to protect the momentum of upward redistribution, which facilitates wealth concentration through austerity, resulting in a dramatic drop in democratic support in many EU member states. Satisfaction with democracy decreased by 45.5% between late 2007 and 2011 in Greece, and by 32.1%, 17.8%, and 16.3% in Spain, Cyprus, and Slovenia respectively. Trust in national parliaments fell by 39.8% in Greece, 29.3% in Spain, 21.9% in Cyprus, and 21% in Slovenia during the same period (Armingeon and Guthmann, 2014: 432). Armingeon and Guthmann found that the imposition of austerity measures from above, which largely neglected national democratic decision-making, contributed to the realization of European people that their national democratic institutions were severely constrained and no longer deserved public support. There is little wonder, then, that "democratizing Europe" has become a popular term and a central theme to numerous social movements, as exemplified by the "Manifesto for the democratization of Europe" (Piketty and Vauchez, 2018) and the "Democracy in Europe Movement 2025 (DiEM 25), which sets 2025 as the date by which Europe must be democratized.

In sum, the more meaningful question than "whether democracy is in the way" and "whether more rather than less democracy is the solution" seems to be *"what kind* of democracy is being practiced throughout the democratic world?" Works by Gilens and Page, Milanovic, Piketty, Mair, Armingeon and Guthmann, and many others seem to suggest that *the* kind of democracy being practiced in advanced democracies today is the *neoliberalized* kind. Neoliberalized democracy is zombie democracy that bears the appearance of inclusivity while performing, in stealth, the function of funneling power, wealth, and resources into the hands of ever fewer individuals. The major tools that have helped to enable the neoliberalization of democracy have been money-sustained dissemination of rhetoric, discourse, persuasion, propaganda, and ideas and beliefs specifically manufactured for the purpose of deceit. With these tools, neoliberalism got to shape democracy by filtering out information and policies useful for the public but detrimental to the rich and powerful, while at the same time promoting information and policies useful for facilitating the interests of the affluent few, even at the expense of the wellbeing of the public.

In *Silent Spring*, Rachael Carson offered glimpses of early stages of the neoliberalization of democracy in the U.S. and its damage on the environment:

> This is an era... dominated by industry, in which the right to make a dollar at whatever cost is seldom challenged. When the public protests, confronted with some obvious evidence of damaging results of pesticide applications, it is fed little tranquilizing pills of half truth. We urgently need an end to these false assurances, to the sugar coating of unpalatable facts. It is the public that is being asked to assume the risks that the insect controllers calculate. The public must decide whether it wishes to continue on the present road, and it can do so only when in full possession of the facts. (1962: 23)

In the introduction to the *40th Anniversary Edition of Silent Spring*, Lear pointed out that the book was about the urgency of changing "how democracies and liberal societies operated so that individuals and groups could question what their governments allowed others to put into the environment." Believing that the federal government was part of the problem, Carson urged her readers to ask "Who Speaks, And Why?" The kind of social revolution that Carson wished to ignite was therefore "a democratic activist movement that would not only question the direction of science and technology but would also demand answers and accountability" (Lear, 2002).

The period featured in *Silent Spring* is the period when the conservation movement in the U.S. was experiencing a dramatic change. Citing Pichot, one of the founding figures of the conservation movement, in stating that "[t]he natural resources must be…. preserved for the benefit of the many, and not merely the profit of the few" (Pinchot, 1910: 466),[4] McConnell considered the conservation movement of the first decades of the 20th century as the realization, in political form, of a delusively simple idea of equality (McConnell, 1954: 467). By the mid-1950s, however, it had become apparent that the goals of conservation had become severely diluted. Some conservation groups appeared to represent selfish special interests that were being denounced in the Progressive era (McConnell, 1954: 467).[5] While in Pichot's time, conservation was the cause of the public interest against particular interests and "the defense of the common heritage against predatory selfishness," by the mid-20th century, "[a]ny decision that will in fact be made will be in terms of the particular set of values held by the administrator or, perhaps, by the particular set of pressures that are brought to bear on him" (McConnell, 1954: 468, 471).

This background against which the story of *Silent Spring* was told gives some hints as to why even though the specific chemical products such as DDT that caused the springs to be silent were banned mainly as a result of the publication of the book, the structure that gave chemical and other corporations their disproportionate influences grew even stronger. By the 1980s it became apparent that "the very private interests which were to be controlled by public servants (acting in the public interest) came themselves to dominate the resource management agencies." These private interests have in effect come to "determine the public interest jointly with those in the employ of the public bureaucracies" (Paehlke, 1988: 295).

Where democracy remains trapped in its neoliberalized form—"more rather than less democracy"—is a guarantee for annihilation. Embracing authoritarianism, however, does not yield a different result for reasons that Passmore, Paehlke, Dryzek, and Stehr have explained. The only options left, therefore, are to sort out how democracy became neoliberalized, the mechanisms through which it took place, and to devise ways to bypass neoliberal controls in practicing democracy so that effective climate mitigation becomes possible and survival becomes attainable, even for the non-rich and non-powerful. The next chapter traces the neoliberalization of democracy to the Keynes-Hayek "debate," demonstrating that the decadence of democracy was the intended consequence of a painstakingly designed economic-political order rather than the mere sum of mindlessness and sloppiness scattered here and there.

Notes

1 Naomi Oreskes emphasizes elsewhere the importance of preserving democracy and grounding climate solutions in technological and political realities to avoid the worst impacts of climate change (Oreskes, 2016: 9–10).

2 Dryzke notes that this judgment is supported by a reading of Paehlke (1988), Paehlke and Torgerson (1990), Walker (1988) and Orr and Hill (1978), Dryzek (1987), 88–109, and Dryzek (1992).
3 For a similar view, see Bocking (2004) 200–4.
4 Cited in McConnell (1954) 466.
5 The Progressive era refers to the period spanning from the late 1890s to the early 1920s in United States history. It was characterized by the booming of social activism aimed at tackling problems that rose from the excesses of the Gilded Age, including industrialization and urbanization. See Nugent (2010).

References

Armingeon, Klaus and Kai Guthmann. (2014). "Democracy in Crisis? The Declining Support for National Democracy in European Countries, 2007–2011," *European Journal of Political Research*. Vol. 53: 423–442.

Bocking, Stephen. (2004). *Nature's Experts—Science, Politics, and The Environment*. New Brunswick, New Jersey: Rutgers University Press.

Caldwell, Lynton K. (1963). "Environment: A New Focus for Public Policy?" *Public Administration Review*. Vol.23(3): 132–139.

Campbell, Peter Scott. (2013). "Democracy v. Concentrated Wealth—In Search of a Louis D. Brandeis Quote," *Green Bag*. Vol. 16, no. 3.

Carson, Rachael. (1962). *Silent Spring*. Greenwich: Fawcett Publications.

Cartledge, Paul. (2016). *Democracy: A Life*. Oxford: Oxford University Press.

Dahl, Robert, A. (1970). *After the Revolution? Authority in a Good Society*. New Haven: Yale University Press.

Dryzek, John S. (1987). *Rational Ecology: Environment and Political Economy*. Oxford: Basil Blackwell.

Dryzek, John S. (1992). "Ecology and Discursive Democracy: Beyond Liberal Capitalism and the Administrative State," *Capitalism, Nature, Socialism*. 3(2), 18–42.

Dryzek, John S. (1996). Strategies of Ecological Democratization, in Lafferty, William M. and James Meadowcroft (eds.). *Democracy and the Environment—Problems and Prospects*. Cheltenham: Edward Elgar. 108–123.

Gilens, Martin and Benjamin I.Page. (2014). "Testing Theories of American Politics: Elites, Interest Groups, and Average Citizens," *Perspectives on Politics*. Vol. 12, No. 3, 564–581.

Gilley, Bruce. (2012). "Authoritarian Environmentalism and China's Response to Climate Change," *Environmental Politics*, Vol. 21, No. 2, 287–307.

Heilbroner, Robert. (1980) [1974]. *An Inquiry into The Human Prospect*. New York: W. W. Norton & Company.

Hobsbawm, Eric. (2007). *Globalisation, Democracy and Terrorism*. London: Little Brown.

Josephson, Paul R. (2005). *Resources Under Regimes: Technology, Environment, and the State*. Cambridge: Harvard University Press.

Krosnick, Jon A. and Bo MacInnis. (2013). "Does the American Public Support Legislation to Reduced Greenhouse Gas Emissions?" *Daedalus*. 142(1): 26–39.

Lear, Linda. (2002). Introduction to the 40th Anniversary Edition of Silent Spring, in *Silent Spring*. Boston:Houghton Mifflin Company.

Mair, Peter. (2013). *Ruling The Void—The Hollowing of Western Democracy*. London: Verso.

McConnell, Grant. (1954). "The Conservation Movement—Past and Present," *Western Political Quarterly*. Vol.7: 463–478.

Milanovic, Branco. (2016). *Global Inequality—A New Approach for the Age of Globalization*. Cambridge, Massachusetts: Harvard University Press.

Nugent, Walter. (2010). *Progressivism—A Very Short Introduction*. Oxford: Oxford University Press.

Ophuls, William. (1977). *Ecology and the Politics of Scarcity—Prologue to a Political Theory of the Steady State*. San Francisco: W. H. Freeman and Company.

Ophuls, William and A. Stephen Boyan, Jr. (1992). *Ecology and the Politics of Scarcity Revisited*. New York: W. H. Freeman and Company.

Oreskes, Naomi. (2016). "FORUM." *Issues in Science and Technology, 32*(3), 9–10.

Oreskes, Naomi and Erik Conway. (2014). *The Collapse of Western Civilization—A View from the Future*. New York: Columbia University Press.

Orr, David W. and Stuart Hill. (1978). "Leviathan, the Open Society, and the Crisis of Ecology," *Western Political Quarterly*. 31. 457–469.

Paehlke, Robert. (1988). "Democracy, Bureaucracy, and Environmentalism," *Environmental Ethics*. 10: 291–308.

Paehlke, Robert. (1996). Environmental Challenges to Democratic Practice, in Lafferty, William M. and James Meadowcroft (eds.). *Democracy and the Environment–Problems and Prospects*. Cheltenham: Edward Elgar. 18–38.

Paehlke, Robert and Douglas Torgerson (eds.) (1990). *Managing Leviathan: Environmental Politics and the Administrative State*. Peterborough, Ontario: Broadview.

Passmore, John. (1974). *Man's Responsibility for Nature*. London: Duckworth.

Pichot, Gifford. (1910). *The Fight for Conservation*. New York: Doubleday Page and Company.

Piketty, Thomas. (2014). *Capital in the Twenty-First Century*. Cambridge, Massachusetts: Harvard University Press.

Piketty, Thomas and Antoine Vauchez. (2018). "Manifesto for the Democratization of Europe." www.socialeurope.eu/manifesto-for-the-democratization-of-europe.

Schattschneider, Emer E. (1960). *The Semi-sovereign People—A Realist's View of Democracy in America*. New York: Wadsworth.

Shearman, David J. C., and Joseph Wayne Smith. (2007). *The Climate Change Challenge and the Failure of Democracy, Politics and the Environment*. Westport, CT: Praeger.

Stehr, Nico. (2016). "Good Climate, Bad Democracy," Accessed February 2016. www.thegwpf.com/nico-stehr-good-climate-bad-democracy.

Stehr, Nico. (2016). "Exceptional Circumstances: Does Climate Change Trump Democracy?" *Issues in Science and Technology*. 32(2), 37–44.

Taylor, Bob Pepperman. (1996). Democracy and Environmental Ethics. In Lafferty, William M. and James Meadowcroft (eds.). *Democracy and the Environment—Problems and Prospects*. Cheltenham: Edward Elgar. 86–107.

Walker, K.J. (1988). The Environmental Crisis: A Critique of Neo-Hobbesian Responses. *Polity*. 21. 67–81.

Witherspoon, Sharon. (1996). Democracy, the Environment and Public Opinion in Western Europe, in William M. Lafferty and James Meadowcroft (eds.), *Democracy and the Environment—Problems and Prospects*. Cheltenham: Edward Elgar, 39–70.

2 "Free" Competition of Ideas?

The most important feature of neoliberalism is that the society that neo-liberals envision, i.e., one with a shrinking place for meaningful democracy, is not one that can be *honestly* communicated to the public. This chapter deals with the role that the so-called "debate" between Hayek and Keynes played in handling this inconvenience and quietly turning the world into a grabber's paradise.

In their book, *Merchants of Doubt—How A Handful of Scientists Obscured the Truth on Issues from Tobacco Smoke to Global Warming*, Oreskes and Conway laid bare the way that the "Tobacco Strategy" was copied and suc-cessfully applied to and worked for an array of industries, from asbestos to fossil fuels, all of which had incentives to discredit science. The strategy entailed disguising corporate financing in the form of "think tanks" and "research institutes" to house ideologically devoted scientists, as well as public relations (PR) gurus, whose task was to undermine science. In this chapter I demonstrate that the reason that the "Tobacco Strategy" has worked so smoothly, and the selling of doubt about science has been so effective, across industries and over time, was fundamentally connected with the existence of the neoliberal economic playbook. The hidden agenda of neoliberalism was no less than to reconstruct the economic order into one that concentrated ever more wealth, resources, and power in the hands of ever fewer people. The original playbook, in other words, was written not by the tobacco industry, but the "masters of the universe" featured in Daniel Stedman Jones's *Masters of the Universe,* which sheds light on the birth of neoliberal politics (Stedman Jones, 2012). The rules according to which the universe must operate were laid down long before the climate issue came on to the scene.

Neoliberalism is largely the result of the triumph of the Hayekian economic view over the Keynesian one. On the face of it, it was a free competition of ideas: Hayek's theory must have been deemed either closer to the economic reality, more logical at the theoretical level, more feasible to policymakers, more popular to the general public, or more capable of standing the test of time. A closer look at the manner in which the Hayekian view eventually replaced Keynesianism as the dominant governing principle in advanced capitalist states since the 1980s, however, reveals that what gave rise to neoliberalism was rather the alignment of Hayek's theory with business interests able and willing to fund "think tanks" and "research institutes" to influence policies that supported the business agenda. In other words, the victory of Hayek in this competition of ideas was not "free" at all. As will be seen, the process through which Hayekian thought was promoted at the inception of the rise of neoliberalism bore strong resemblance to the "Tobacco Strategy" described in *Merchants of Doubt*. The consequences of this success, however, were much more profound and far-reaching for all species on the planet. It amounted to an unwritten constitution for economic *as well as* political activities in the capitalist world, laying the foundation for an economic and political order that explains the ease with which the "Tobacco Strategy" worked smoothly and successfully time and again.

The debate between Keynes and Hayek was about much more than how the economy works. Central to the debate, which lingers to this day, is the understanding and perception of democracy and the relationship between economics and politics in a democracy. Keynes, with his demand-side management, considered full employment, reasonable wage levels, and a welfare state supported by public spending key to the health of the economy. Hayek, by contrast, took Keynes' concern with the unemployed and disadvantaged as an indication that Keynes was too "political" to even be qualified as an economist (Cockett, 1995: 46). Where Keynes' macroeconomic theory saw a crucial role for government in managing aggregate demand, Hayek saw such government interference as sowing the seeds for bigger problems down the road. The most crucial difference between Keynes and Hayek, however, was not their beliefs in how the economy worked "objectively"—both believed that the market had its own logic, and both recognized that some minimal social protections must be provided by the government—but what a free society would entail.

Like the word "ecology," the word "economy" has its origin in the Greek word "Oikos," meaning "family" or "house." For Aristotle, "economics" was the science of "household management." By treating social and political members of an economy as indispensable, Keynes' theory kept the meaning of "economics" intact. While treating its members as indispensable seems to be the only sensible way of building an economic theory for a free society, given that "society" is by definition about *everyone* in it, Hayek's theory managed to treat economy as detached from, and superior to, society, rendering the exclusion of some members of society not only tolerable but desirable for the sake of the economy. The exclusionary nature of Hayek's theory played a significant part in attracting sympathetic business funders who would vigorously disseminate and promote his theory, beginning in the late 1940s, and help to shape his theory to even better reflect their interests. This neoliberal prototype for information and idea dissemination would, in the decades to come, serve as the abstract "hub" for other neoliberal "dispatchers," tasked variously to serve the interests of tobacco, toxic chemicals, genetically modified organisms, welfare "reforms," tax "reforms," and climate denial.

Readily seeing some members of society as dispensable, Hayek's economic theory held that even in the worst recessions, the economy would find its way—through the market-determined rise or fall of wages and interest rates—back to equilibrium at full employment, as long as it was left to itself; Hayek saw little positive effect in government interference in the economy.[1] Initially Keynes also expected runaway unemployment, like that which occurred in Britain in the 1920s, to eventually fall back to "normal." When, however, instead of resettling on a full employment equilibrium, the new equilibrium rested and stayed at a stubborn "underemployment" level, Keynes began to look for possible blind spots in classical economic theories. In doing so, Keynes revolutionized economics by bringing attention to aggregate demand. Arguing that the key lesson from the 1929–32 recession was there were no automatic forces in a market economy to induce a recovery, Keynes advocated the government's active role in rekindling the animal spirit in businessmen and bringing aggregate demand back up to a level where all willing workers were employed. Lacking this government-initiated stimulus, the economy could go on shrinking until it reached, and thereafter remained stuck in, the "under-employment equilibrium" (Skidelsky, 1996: 25; 2011; Galbraith, 1965). Hence, for Keynes, full employment, reasonable wage levels, and a welfare state supported by public spending were key to the health of the economy. Only when aggregate demand was maintained by ensuring that those most likely to spend rather than save their income (i.e., wage earners rather than capitalists or suppliers), would the economy become both dynamic and stable. The corollary of Keynes' observation was that the adept usage of fiscal and monetary instruments by government was essential. The role of the government was hence to increase public spending and to stimulate the economy in a slump, while increasing taxes in the upturns. Only after the government had learned to manage the economy by countering the natural tendencies at the micro level, could macroeconomic stability be possible.

Climate scientists concerned about human-induced global warming do not dispute that Planet Earth is subject to life cycles, nor that the conditions of the planet cannot forever be suitable for humans to live in. In the same way, Keynes did not dispute that market works according to a logic of its own and that, if left to itself, the adjustment of interest rates and wages may settle at a reasonable level "in the long run." Yet for an economist who saw members of the society as indispensable:

> [t]he long run is a misleading guide to current affairs. In the long run we are all dead. Economists set themselves too easy, too useless a task if in tempestuous seasons they can only tell us that when the storm is past the ocean is flat again. (1923: 80)

Unemployment is more than just a temporary mismatch of production and atomic, faceless labors. If the "natural law" of market stipulates a ten-year high unemployment period, the unemployed do not simply bounce right back into the work force in the 11th year after the slump has run its course. The survival, health, self-esteem, skills, knowledge, and social network of the unemployed, and the life chances of their children, become drastically and permanently altered when unemployment lasts for extended periods. For economists subscribing to the subordination of society to economy, such casualties are not only acceptable but also imperative. This subordination of society to the economy was exactly the potential that the sympathetic business funders saw in Hayek's theory. Subordinating society to economy necessarily excludes. Exclusion by definition leaves more resources—resources that were otherwise there for the subsequently excluded to enjoy—for the included. As long as the subordination could obtain the constitutional status and set the perimeter for the rule of law of the system, the only possible consequence would be the exclusion of ever more individuals from resources, which would remain at the disposal of ever fewer individuals. Already a parallel can be found in the essence of Hayek's theory and climate change: a beautiful world is not only possible but probable in the climate-changed planet: a sufficient amount of population being wiped out to make the planet more livable for the few. Climate change is a catastrophic problem only if one insists on including everyone. If Hayek's exclusionary economic theory was accepted, how can the exclusion of some human beings be a problem when those at the top still get to build a climate-resistant life? So what if the climate catastrophe tends to hit the least resourceful and the least responsible for climate change the hardest? If exclusion is the accepted norm in the economic realm, why suddenly become agitated by exclusion executed through the form of climate disasters?

The Keynes-Hayek debate was important to the discussion of climate change not because one economic theory was "right" and the other "wrong," or that had we followed the "right" economic theory, catastrophic climate change could never have happened. The Keynes-Hayek debate was a critical juncture because the path envisioned by Keynes, being inclusive in nature, would have left many possibilities open, including moving on from Keynesianism to even more inclusive economic paradigms. By contrast, Hayekian exclusionary reasoning, particularly after it was seized, expanded, consolidated, and radicalized by followers including the likes of Charles Koch, foreclosed policy options that would have served to reverse the exclusionary momentum. For Keynes, the key question was "whether we are prepared to move out of the nineteenth-century *Laissez-faire* state into… a system where we can act as an organized community for common purposes and to promote economic and social justice" (1939: 123).[2] This ideal of Keynes is inherently compatible not only with democracy but also with climate mitigation. In contrast, Hayek saw the emphasis on economic and social justice as an abandonment of economic freedom. He was exasperated by the social order implied by Keynes' vision—that is, the phenomenon that "the beliefs of the great majority on what was right and proper were allowed to bar the way of the individual innovator." The result of this dominant view, he claimed, was the tragic undertaking to "dispense with the forces which produced unforeseen results and to replace the impersonal and anonymous mechanism of the market by collective and 'conscious' direction of all social forces to deliberately chosen goals" (2007 [1944]:70, 73). To mock the notion of "freedom from necessity" found in this horrible "collectivism," Hayek pointed out that:

> although the promises of this new freedom were often coupled with irresponsible promises of a great increase in material wealth in a socialist society; it was not from such an *absolute conquest of the niggardliness of nature that economic freedom was expected.* What the promise really amounted to was that the great existing disparities in the range of choice of different people were to disappear. (2007 [1944]: 78; emphasis added)

The entitlement of the rich few to unlimited extraction from "the niggardliness of nature" and the immorality of a more equal distribution of wealth in Hayek's theory bode ill with our current task of climate mitigation.

Hayek's distaste for collectivist experiments/systems such as fascism and communism can explain his peculiar understanding of a democratic system in which individualism remains the supreme guiding principle, even to the extent of entailing systemic and collective grabbing from the already underprivileged members of society.[3] His exclusionary theory appeared acceptable not only because it seemed objective, independent, and detached, like natural science, but also because the underprivileged who were likely to be excluded were not named: *everyone* had equal opportunity to ensure that he would not be the unfortunate sacrifice.[4] By mistaking an inclusive democratic system for collectivism like fascism or communism, Hayek's intrinsically exclusionary theory became music to the ears of wealthy businessmen, who understandably seized the opportunity to promote his theory.

Had the main contender to Keynes' theory been the ecology-conscious Karl Polanyi instead of Hayek, the climate crisis might not have had a place in the 21st century. This, of course, is just a fantasy. The appeal of Hayek's theory to big businesses guaranteed opulent funding for its dissemination and hence its prolonged relevance. The great irony in the "free" competition of ideas was that, the "world-view" that won the race—the idea that stressed the importance of "free" competition—had neither the merit of intellectual superiority nor theoretical validity. Instead, it owed its success largely to the huge piles of promotional money analogous to campaign contribution or PR fees. Hence, to think of neoliberalism as an "ideology" wrongly elevates it to a level to which it does not belong. Instead, neoliberalism is better recognized as a bag of money-sustained sophisticated tools and salesman's techniques, which are useful for double-talk aimed at disguising the exclusionary nature of the distributive order that neoliberals determinedly attempt to normalize.

Before describing this ironic "free" competition of ideas, I will first focus on Polanyi's analysis, which surpasses both Keynes and Hayek in terms of scope and depth, particularly from the perspective of current conditions. While the conclusion of *The Great Transformation* may lead some people to see ethics, morality, and altruism as the backbone of Polanyi's theory, it is in fact logic, sound reasoning, and solid historical evidence that filled the book. In fact, had the terms "market" and "rationalist" not been reserved, starting in the 1980s, to denote something quite different from their original meanings, characterizing Polanyi as a "pro-market" "rationalist" would not be inappropriate. Hayek's theory is informed by the belief that excluding those at the bottom of society is acceptable or even desirable, given the sacrosanctity of the self-regulating market. In contrast, while recognizing the self-regulating feature of the market, Keynes' theory is informed by his belief that the well-being of *all* in society must be taken into consideration *in time* (as in the long run we are all dead), thus validating the necessity of government intervention. Polanyi's socioeconomic theory is different from both in demonstrating that the so-called self-regulating market did not and never will work for a prolonged period of time, precisely because of its exclusionary effect.

History has shown that the only kind of market that can operate smoothly for a prolonged period of time is the kind that is embedded in society, serving its need, and therefore which is regulated, as was the case in human history up until the end of the 18th century. It was crucial to recognize that the nature of such a market was limited and non-expansionary, and that isolated markets did not automatically expand into a market economy, nor did regulated markets simply evolve into a self-regulating market. Such developments were rather "the effect of highly artificial stimulants administered to the body social" (1944: 60). The limited and non-expanding market, embedded in and serving the needs of society, set highly inconvenient boundaries for individuals eager to maximize their private wealth. With coercive power, such individuals helped to rush the transition from feudal-agricultural to market-industrial society, denying the time needed for society to prepare and adjust to industrialization.

The new system that hinged on the disembedding of the market from society would not have lasted for a century, or contributed so strongly to the forces that led to two devastating World Wars (not to mention subsequent conflicts, including the Cold War), had thinkers of the time not developed positivist and utilitarian theories to fit the nascent reality into the scheme of philosophy and theology in order to assimilate the disruptive socioeconomic order with human meanings (1944: 87). Classical economics arose under similar conditions to the dubious "science" of climate denial. As genuine climate science began to make clear the consequences of fossil fuel combustion, those who had profited from resource extraction needed to do two things simultaneously: to maintain their position (by continuing to extract and sell fossil fuels) and to legitimize their actions (by making their activities appear less destructive, negligent, and self-serving). Climate denial science served this purpose, casting doubt on how serious, indiscriminate, and destructive fossil fuel extraction and combustion have been. In the same way, the "science" of classical economics not only concealed, but even legitimized the wanton social destruction of the enclosure of the commons and the forced eviction of people from their land, portraying it as a sort of "natural law," rather than simply serving the interests of the industrial and social elite. Classical economists such as Ricardo and Malthus were thus critical in "administering the artificial stimulants to the body social."

The expansive, omnipotent, and self-regulating One Big Market (1944: 75), the operating logic of which the classical economists claimed to have discovered, could exist only in fiction. The most important "commodities" featured in their theories—land, labor, and money—were not real commodities, namely, objects produced for the purpose of being sold on the market and subject to the supply-and-demand mechanism interacting with price.

> For the alleged commodity "labor power" cannot be shoved about, used indiscriminately, or even left unused, without affecting also the human individual who happens to be the bearer of this peculiar commodity. In disposing of a man's labor power the system would, incidentally, dispose of the physical, psychological, and moral entity "man" attached to the tag. (1944: 76)

As to land, it is "only another name for nature, which is not produced by man" (1944: 75).

> What we call land is an element of nature inextricably interwoven with man's institutions. To isolate it and form a market for it was perhaps the weirdest of all the undertakings of our ancestors. Traditionally, land and labor are not separated; labor forms part of life, land remains part of nature, life and nature form an articulate whole. (1944: 187)

By pretending that land *was* commodity, "[n]ature would be reduced to its elements, neighborhoods and landscapes defiled, rivers polluted, military safety jeopardized, the power to produce food and raw materials destroyed" (1944: 76). George Orwell's critique on land-owning was equally illuminating:

> … consider how the so-called owners of the land got hold of it. They simply seized it by force, afterwards hiring lawyers to provide them with title-deeds. In the case of the enclosure of the common lands, which was going on from about 1600 to 1850, the land-grabbers did not even have the excuse of being foreign conquerors; they were quite frankly taking the heritage of their own countrymen, upon no sort of pretext except that they had the power to do so.[5] (Orwell, 1996 [1944]: 207)

Finally, money, the most fiercely pursued commodity of our time, is, according to Polanyi, "merely a token of purchasing power which, as a rule, is not produced at all, but comes into being through the mechanism of banking or state finance" (Polanyi, 1944: 75).

The pretense that these "fictitious commodities" must be treated as if they were non-fictitious commodities, and the insistence on organizing society as if they were real commodities, could only result in the demolition of society. "[N]o society could stand the effects of such a system of crude fictions even for the shortest stretch of time unless its human and natural substance as well as its business organization was protected against the ravages of this satanic mill" (1944: 76). It should have been common sense that nature cannot possibly be forced to subordinate to society, and society in turn to market. Instead, esoteric and science-like economic theories stepped in to "correct" the coarse intuition of ordinary people in order to trap human society in the market-society and market-nature groove. Thus, "[t]he road to the free market was opened and kept open by an enormous increase in continuous, centrally organized and controlled interventions," hence Polanyi's famous remark that: "[l]aissez-faire was planned" (1944: 146).

Given that the arrogant defiance of the order among nature, society, and market was meant to institutionalize the exclusion and the legitimization of resource-grabbing, the resulting, cumulative social disruption could end only in class and racial tensions and, ultimately, conflict.

> For a century the dynamics of modern society was governed by a double movement: the market expanded continuously but this movement was met by a countermovement checking the expansion in definite directions. Vital though such a countermovement was for the protection of society, in the last analysis it was incompatible with the self-regulation of the market, and thus with the market system itself. (1944: 136)

The Great Transformation not only provides crucial information and analysis about the intellectual linage from which Hayek's theory descended, but also offers clues for understanding the astonishing easiness with which a host of denialism from tobacco to climate has spread and successfully intervened in public policymaking. The right winds have to already exist to disseminate science-like, gains-driven discourses. Industry-serving, policy-informing classical economics was the first identifiable jet stream to feed the favorable winds necessary to reverse the nature-society-market order, which formed the popular world-view. In our time, it is neoliberalism, with improved PR tools and the aid of democracy, that has risen to prominence, surpassing even classical economics, in administering in the body social a highly manipulated market order that crushes both society and nature. Against this background, climate change has no chance of being tackled with necessary profundity, unless it is treated as a sub-issue or residue of economic order. Given this critically important background, climate change has to be understood in the same way as battles over not only tobacco and pesticides, but also topics such as income distribution, labor law, trade policy, and tax systems. These have historically been the areas that neoliberalism attacked first, striking any part of the system where society was viewed as inclusive and where, by extension, no part of that inclusive society could be seen as dispensable.

A momentous upward wealth re-distribution serves as a turbo-charged engine precipitating the destruction of nature. Neoliberalism is not only about the upward re-distribution of wealth, but also about *expanding* the total sum of private wealth in the world. This expansion comes from either nothing, as is achieved through the magical financial market, or something, which entails the grabbing of what belonged to the commons or nature. As nature gets closer to depletion under the neoliberal system and ever fewer resources are left in the world to seize, it becomes harder to attend to manners. Polanyi's interpretation of history goes a long way to explain the increasingly blatant lack of civility that we witness in the 21th century when the system teeters towards an end.

Hayek lambasted the development where "the beliefs of the great majority on what was right and proper were allowed to bar the way of the individual innovator" (1944: 70) and lumped such beliefs with "collectivism." In the same way, modern neoliberal elites vilify those who question the economic and environmental justice of the current system as anti-democratic populists or environmental/ecological terrorists. Such defamations, reminiscent of the assertion by the 19th-century economic liberals that "anti-laissez-faire legislation was the result of purposeful action on the part of the opponents of liberal principles" (Polanyi, 1944: 147), are not only anti-democratic but also anti-logic and anti-history. As Polanyi's analysis showed, the rise of a countermovement needed neither conspiracy nor collectivism. It rose, rather, as a spontaneous, natural, and necessary consequence of the disembedded market. Hence, while "[l]aissez-faire was planned; planning was not" (1944: 146).

To situate Polanyi's theory back into the Hayek-Keynes debate, although the main difference between Hayek and Keynes is the legitimacy of state intervention in the market, Polanyi's theory questions the very notion of "intervention," because the mythical self-regulating market that the state is supposedly encouraged or discouraged to intervene can exist only for a prolonged period of time in fiction. To maintain a truly sustainable market system, Polanyi saw no alternative to re-embedding the market in society through regulation. From this perspective, Polanyi is the ultimate "pro-marketer," given his devoted concern to the long-term health of the market.

Seen from today's climate-altered world with rising sea-levels, droughts, floods, and war, Polanyi's theory was doubtlessly the most insightful and prescient of the three discussed above with its treatment of nature as an uncompromising force to which society must be subordinated. The attention that Polanyi's theory received, however, was comparable neither to Keynes nor to Hayek. While John Ruggie had seemingly popularized Polanyi's theory by coining the term "embedded liberalism" in his portrait of the postwar economic order as a reconciliation between vibrant international free trade and domestic social compensation (Ruggie, 1982), such a characterization of the Bretton Woods system is seen by others as a mythologization of a design that enlisted, rather than suppressed, private American finance that sought to further disembed the market from society (Appleton, 2016).

Polanyi's reading of history gave the social and industrial elites who were poised to manipulate the market few incentives to invest resources in his theory. The opposite was true for Hayek. His view placed him "close to the heart of all those American businessmen, politicians, and others who had never been reconciled to Roosevelt's policies" (Stedman Jones, 2012: 63). In the end, the critical historical juncture involved only Keynes and Hayek. Keynes' theory might have been less thoughtful sociologically and more limited in its scope in comparison with Polanyi's theory, but with its emphasis on the role of the state it would nonetheless have allowed society to adapt to newly obtained scientific knowledge such as climate change. Such agility was exactly what the neoliberal project targeted and sought to eliminate.

We now live in a world far from the one envisioned by Keynes, in which we "act as an organized community for common purposes and to promote economic and social justice," largely because Hayekian ideas have so closely aligned with the interests of the affluent. Even though Keynes died in 1946 and did not see his theory fall out of favor, he was fully aware how easily ideas of individualism could win over the hearts of liberal societies, because "[t]hese ideas accorded with the practical notions of conservatives and of lawyers. They furnished a satisfactory intellectual foundation to the rights of property and to the liberty of the individual in possession to do what he liked with himself and with his own" (1926 [1924]: 272–3). Hence, noted Keynes, individualism and *laissez-faire*, "in spite of their deep roots in the political and moral philosophies of the late eighteenth and early nineteenth centuries," could not have "secured their lasting hold over the conduct of public affairs, if it had not been for their conformity with the needs and wishes of the business world of the day" (1926 [1924]: 283).[6] Although this insight came from his 1924 Sidney Ball Lecture on "The End of Laissez-Faire," it might as well have been the script for the rise of neoliberalism in the decades that followed.

One of the earliest American business funders for the spreading of Hayek's ideas was the Rockefeller Foundation. Starting in the late 1920s, the Rockefeller Foundation provided substantial funding to both the London School of Economics (LSE), where Hayek and Lionel Robbins were based, and the Institut Universitaire des Hautes Etudes Internationales in Geneva, where William Rappard, Wilhem Röpke, Louise Rougier, and Ludwig von Mises were based. What Cockett called the "economic counter-revolution" against Keynesianism began with these individuals who devoted their intellectual strength to developing "an articulate and coherent critique of 'planning', collectivism and Keynesianism" (Cockett, 1995: 55).

An astounding amount of work came out of this group, the content of which warranted continued and expanded funding from the business world. In *Bureaucracy*, published in 1944, for instance, Hayek's mentor, von Mises, portrayed the government as the "guardianship of a gigantic apparatus of compulsion and coercion" that stood in the way of freedom, private initiative, individual responsibility, and democracy (1944: iiv).[7] There was no remedy for the detrimental effects of government bureaucracies on the public interest, other than replacing the system with one whose operational logic was governed by private profit. Rather than government bureaucracies, argued von Mises, corporations were the drivers of social and economic progress: "The great businessman is he who produces more, better, and cheaper goods, who, as a pioneer of progress, presents his fellow men with commodities and services hitherto unknown to them or beyond their means." The great businessman "embodies in his person the restless dynamism and progressivism inherent in capitalism and free enterprise" (1944: 13).

Doubtlessly, such arguments were music to the ears of wealthy businessmen, whose interests in funding this group of economists were further elevated by the publication of Hayek's *The Road to Serfdom* in 1944. Tossing out of the window his accusation about Keynes being too political, Hayek began the book by confessing that it was "a political book." The function that Hayek hoped the book would perform was quite evident. Setting out to make as big an impact on public opinion as possible, not only did Hayek write it in a way that appeared detached and non-partisan, but he was also assiduous in making his message as easily accessible to a wider public as possible. He distilled his key arguments into the 12 "Points from the Book" and printed on the back of the yellow dust-cover (Cockett, 1995: 80, 87). The "Points from the Book," which had the tinge of the playbooks of Tobacco Strategy described in *Merchants of Doubt*, included the following:

- Totalitarianism is the new word we have adopted to describe the unexpected but nevertheless inseparable manifestations of what in theory we call socialism.
- In a planned system we cannot confine collective action to the tasks on which we agree, but are forced to produce agreement on everything in order that any action can be taken at all.
- The economic freedom which is the prerequisite of any other freedom cannot be the freedom from economic care which the socialists promise us and which can be obtained only by relieving the individual at the same time of the necessity and of the power of choice: it must be the freedom of economic activity which, with the right of choice, inevitably also carries the risk and the responsibility of that right. (Cockett, 1995: 81)

Although the book was a big success, the seemingly mild criticisms that came from Orwell and Keynes might have proven fatal to Hayek's thesis, had the Keynes-Hayek debate remained intellectual, instead of becoming a competition of funding and PR campaigning. Orwell agreed with Hayek that collectivism was not inherently democratic, but gave to "a tyrannical minority such powers as the Spanish inquisition never dreamed of." However, Orwell believed that Hayek "does not see, or will not admit, that a return to 'free' competition means for the great mass of people a tyranny probably worse, because more irresponsible, than that of the State. The trouble with competitions is that somebody wins them." As a consequence, "State regimentation" may be better than "slumps and unemployment" for most people (Orwell, 1944). As to Keynes, his most important comment on the book was over Hayek's failure to explain where the line between free enterprise and planning should be drawn. Agreeing with "virtually the whole of [the book]," Keynes said:

> [y]ou admit here and there that it is a question of knowing where to draw the line. You agree that the line has to be drawn somewhere...But you give us no guidance whatever as to where to draw it. In a sense this is shirking the practical issue.... [Y]ou greatly under-estimate the practicability of the middle course. But as soon as you admit that the extreme is not possible, and that a line had to be drawn, you are, on your own argument, done for since you are trying to persuade us that as soon as one moves an inch in the planned direction you are necessarily launched on the slippery path which will lead you in due course over the precipice. I should therefore conclude your theme rather differently. I should say that what we want is not no planning, or even less planning, indeed I should say that we almost certainly want more. But the planning should take place in a community in which as many people as possible, both leaders and followers, wholly share your moral position.... But the curse is that there is also an important section who could almost be said to want planning not in order to enjoy its fruits but because morally they hold ideas exactly the opposite of yours, and wish to serve not God but the devil. (Keynes, 1944, quoted in Cockett, 1995: 89–90)

What both Orwell and Keynes put their fingers on was the problem of exclusivity in Hayek's imagination of democracy. Already in 1924 Keynes explained where economic theories relying on individualism went wrong: "Economists, like other scientists, have chosen the hypothesis from which they set out, and which they offer to beginners, because it is the simplest, and not because it is the nearest to the facts." Compounded with this need for simplification was the fact that "they have been biased by the traditions of the subject," propelling them to assume a state of affairs

> where the ideal distribution of productive resources can be brought about through individuals acting independently by the method of trial and error in such a way that those individuals who move in the right

direction will destroy by competition those who move in the wrong direction. This implies that there must be no mercy or protection for those who embark their capital or their labour in the wrong direction. (Keynes, 1926 [1924]: 282)

The assumption that unhindered natural selection leads to progress

is a method of bringing the most successful profit-makers to the top by a ruthless struggle for survival, which selects the most efficient by the bankruptcy of the less efficient. It does not count the cost of the struggle, but looks only to the benefits of the final result which are assumed to be permanent.... Profit accrues, under *laissez-faire*, to the individual who, whether by skill or good fortune, is found with his productive resources in the right place at the right time. A system which allows the skillful or fortunate individual to reap the whole fruits of this conjuncture evidently offers an immense incentive to the practice of the art of being in the right place at the right time. Thus one of the most powerful of human motives, namely, the love of money, is harnessed to the task of distributing economic resources in the way best calculated to increase wealth. (Keynes, 1926 [1924]: 283)

It does not take much effort to see why the privileged few in society saw strength in Hayek's theory and flaws in Keynes' and flocked, with opulent funding, to the former. Through the neoliberal art of exclusion—the ability to divert people's attention from recognizing the (high) risk of being unfairly treated to focusing instead on the "equal" (yet extremely slim) opportunity of climbing to the top—the neoliberal movement succeeded in persuading a large majority of voters that they were included as beneficiaries of neoliberal policies, which in reality excluded them not only from a fair share of their economic contribution, but also from meaningful political participation.

It is exactly because there are two sides to the coin of exclusion that wealthy businessmen were driven to help to defend Hayek's theory. The publication of *The Road to Serfdom*, particularly the publication of the condensed version of it by *Reader's Digest* in 1945, linked up like-minded intellectuals and businessmen from Geneva, London, and Chicago. Dr. Albert Hunold, a Swiss businessman involved in the work of the Institut d'Etudes Internationales at Geneva, invited Hayek to the University of Zürich and brought him into the circle of a group of Swiss industrialists and bankers (Cockett, 1995, 102–3). With strong signals of financial backing from rich businessmen, Hayek began to lay the groundwork for the founding of a society whose goal was to "enlist the support of the best minds in formulating a programme which has a chance of gaining general support" (Hayek, 1947).[8]

The now famous Mont Pelerin Society (MPS), which would serve as "a focus for scholars to carry out the long-term task of converting the next generation of intellectuals to a creed of liberalism that was then largely discredited" (Cockett, 1995, 104), held its inaugural conference at Mont Pelerin in April 1947. The early meetings of the Society were funded by European and American sources alike. In Europe, Hunold raised money from Swiss sources to cover the costs of accommodation and travel for European participants (Hartwell, 1995: 26). Sir Alfred Suenson-Taylor, a friend of Hunold and Chairman of London and Manchester Assurance from 1953 to 1961, cooperated with Hunold to fund the meetings, while providing welcome links to a network of wealthy anti-government City financiers. He further negotiated with the Bank of England to unlock funds for the British delegates to travel to early MPS meetings (Cockett, 1995: 108; Shaxson, 2012: 84). The strong links between the MPS and the City of London continued with Sir Alfred Suenson-Taylor via Lord Grantchester, chairman of a major insurance company in the City of London (Shaxson, 2012: 84). With regard to participants from the United States, Hayek entrusted the detailed arrangements to Leonard Read, the general manager of the Los Angeles Chamber of Commerce and founder of the Foundation for Economic Education, which was supported by leading businessmen. To finance the travel of the American participants, including von Mises, Milton Friedman, George Stigler, Henry Hazlitt, and Leonard Read, H.W. Luhnow of the William Volker Charities Trust set up by an American millionaire, William Volker, committed funding (Hartwell, 1995: 26, 33; Stedman Jones, 2012: 155; Cockett, 1995: 108). These wealthy businessmen provided the venture capital for the neoliberal stealth revolution, and the MPS, which served for the coming decades as the research and development (R&D) hub, succeeded in spawning worldwide manufacturing and sales branches.

The opening paragraph of the Statement of Aims of MPS is eerily relevant and insightful for the current world, only with the precipice laid in the opposite direction:

> The central values of civilization are in danger. Over large stretches of the Earth's surface the essential conditions of human dignity and freedom have already disappeared. In others they are under constant menace from the development of current tendencies of policy. The position of the individual and the voluntary group are progressively undermined by extensions of arbitrary power. Even that most precious possession of Western Man, freedom of thought and expression, is threatened by the spread of creeds which, claiming the privilege of tolerance when in the position of a minority, seek only to establish a position of power in which they can suppress and obliterate all views but their own. (The Mont Pelerin Society, 1947)

Eager for his ideas to seep into society through those able to shape public opinion, Hayek arrived at a very expansive definition of intellectuals, or what he called "professional secondhand dealers in ideas," which readers of *Merchants of Doubt* would find familiar. Noting that the "character of the process by which the views of the intellectuals influence the politics of tomorrow is... of much more than academic interest," Hayek emphasized that the typical intellectual, or the secondhand dealers in ideas, need not be an original thinker, nor to "possess special knowledge of anything in particular, nor need he even be particularly intelligent, to perform his role as intermediary in the spreading of ideas." Rather, what qualifies a professional secondhand dealer for his job is "the wide range of subjects on which he can readily talk and write, and a position or habits through which he becomes acquainted with new ideas sooner than those to whom he addresses himself" (1960 [1949]: 372). Hence, while professional men and technicians such as doctors and scientists are listened to with respect, owing to their expert knowledge of their own subjects, and can naturally become carriers of new ideas outside their own field, journalists, teachers, ministers, lecturers, publicists, radio commentators, writers of fiction, cartoonists and artists, too, belong to the same class. Although "usually amateurs so far as the substance of what they convey is concerned," they can nonetheless be "masters of the technique of conveying ideas." After all, outside one's specialized field, everyone is an ordinary man dependent for information and instruction on this type of "intellectuals" who "make it their job to keep abreast of opinion." It is the keen observation that "there is little that the ordinary man of today learns about events or ideas except through the medium of this class" that can best explain Hayek's successful business model for idea dissemination (1960 [1949]: 372).

Hayek's vision about secondhand dealers in ideas must be as enlightening to readers of *Merchants of Doubt* as it was galvanizing for those who stood to gain by planting his ideas into the operational logic of the political system. Across the Atlantic, a web of secondhand dealers in ideas or what Daniel Stedman Jones termed "ideological entrepreneurs," consisting of sympathetic business funders, intelligentsia, PR experts, and journalists, started to take shape in the 1950s to not only disseminate but also to revise—towards an even more extreme *laissez-faire* direction—Hayek's theory. With the R&D branch, i.e., the MPS, already in place, the sympathetic business funders went on to establish think tanks functioning as the manufacture and sales branch of the neoliberal enterprise. Together with the MPS, the Institute of Economic Affairs (IEA, founded in 1955 in Britain), the Institute for Humane Studies (IHS, 1961, in the U.S.), the Heritage Foundation (1973, U.S.),[9] the Centre for Policy Studies (1974, Britain), the Fraser Institute (1974, Canada), the Cato Institute (1977, U.S.), the Manhattan Institute for Policy Research (1977, U.S.), the Adam Smith Institute (1977, Britain), and the Pacific Institute for Public Policy (1979, U.S.) played critical roles in spreading neoliberal ideas.

The individuals who ran these policy institutes were the ideological entrepreneurs who made neoliberal thought accessible.... They helped turn neoliberal thought into a neoliberal political program. They hustled to establish a media presence by raising their profile among sympathetic journalists, and to secure financial robustness for their organizations, and they fought for influence in the political process through the powerful promotion of free markets.... To a large extent this network was held together by the Mont Pelerin Society." (Stedman Jones, 2012: 135)

Among the sympathetic business funders behind these neoliberal enterprises across the continents were Jasper Crane of DuPont Chemicals, William Volker of the picture framing company, H. W. Luhnow of William Volker Fund, Lawrence Fertig of a New York advertising and marketing firm Lawrence Fertig & Co., Antony Fisher of Buxted Chickens, Lemuel Ricketts Boulwere of General Electric, Coors Brewing Company, David Goodrich of Goodrich and Co., Pat Boyle of Canadian forestry company MacMillan Bloedel Limited, Charles White of Republic Oil Corporation, and Donaldson Brown of General Motors, as well as the Shell Oil Company (Stedman Jones, 2012, chapter 4; Cockett, 1995, chapters 3, 4 and 8).

Within this rich men's list, the most catalytic in spreading Hayek's theory, and in molding our world into the shape it is today, was Antony Fisher. He became a pious believer in Hayek's theory after reading the condensed version of *The Road to Serfdom* in the *Reader's Digest* in 1945. Having made his fortune by bringing battery-caged factory chicken farming from the U.S. to the U.K, Fisher set out to establish the IEA, which was formally founded in 1955 and would in the following decades wield deep and far-reaching influence all over the world. From the very start, the IEA was pitched at academics, writers, journalists, broadcasters, and teachers—that is, the secondhand dealers who dictated the belief system of the public. Both Fisher and the co-founder of the IEA, Oliver Smedley, an English businessman, recognized the importance of maintaining the image of the institution as neutral, independent, and objective. In a letter to Fisher, Smedley emphasized that it was imperative that "we should give no indication in our literature that we are working to educate the Public along certain lines which might be interpreted as having a political bias." He explained that "if we said openly that we were re-teaching the economics of the free-market, it might enable our enemies to question the charitableness of our motives" (Cockett, 1995: 131).

The IEA even sought authors whose views were not automatically identified with those of the IEA "but whose basic love of liberty enables them to come most of the way with us on particular subjects... This infiltration in reverse can prove most effective" (Seldon, 1959; quoted in Cockett, 1995: 143). By maintaining the image of independence and non-partisanship, the IEA would be "the artillery firing the shells (ideas). Some would land on target (the intellectuals), while others might miss." The Institute itself, however, would "never be the infantry engaged in short-term, face-to-face grappling with the enemy. Rather, its artillery barrage would clear the way for others to do the work of the infantry later on" (Blundell, 1987: 32).

The IEA's skillful secondhand dealing was used to demonstrate not only the efficacy of economic liberalism but also the applicability of the principles of the free market to *all* areas of society, from the telephone service, to the Welfare State, to education (Cockett, 1995: 142–3). The IEA also set out to transform what university students would learn inside and outside the classroom. Among the students strongly influenced by the IEA was Madsen Pirie, who later set up the Adam Smith Institute in London in 1976. According to him, it was impossible for a British student not to be aware of IEA's work and its ideas. No matter what the student was taught, "he will inevitably encounter the IEA's publications and research documents, and will realize that there is a solid body of academically respectable opinion oriented towards the consumer rather than the central planner." Precisely because "[o]ne generation of students is the next generation of teachers," "the IEA is so well represented among the younger economists now working their way through University departments." As a result, the IEA not only changed the background intellectual climate of the economic debate; it put arguments "directly into the mouths of interested parties" (Pirie, 1975; Cockett, 1995: 190–1). Even the neoliberal legend Milton Friedman[10] once commented that "without the IEA, I doubt very much whether there would have been a Thatcherite revolution" (Cockett, 1995: 158). Ironically befitting to the theme of the current volume, Margaret Thatcher's famous slogan, "There is No Alternative," came from an essay titled "A Changed Climate" by the historian of the Conservative Party, Lord Blake (Cockett, 1995: 217). Analyzing the circumstances in which the IEA's counter-revolution had come of age, Lord Blake wrote, "The Trendy Lefties who were the apotheosis of the 1960s and early 1970s look like dinosaurs." To tidy up the mess left by them, argued Lord Blake, the Conservatives had a lot of work to do.

> Getting "government off our backs," "setting the people free," "a property-owning democracy"—these are admirable Conservative objectives, but they will require hard work, determination and toughness if they are to be achieved. Yet *there is no real alternative...* Nothing less will suffice than a major reversal of the trends which ever since 1945 Labor has promoted. (Blake, 1976: 3 and 12, emphasis added)

Despite the massive influence that the IEA already had on the policy world, Fisher's ambition, however, did not stop at the IEA. In a letter to Hayek written after hearing of Hayek's Nobel Prize in Economics in 1974, Fisher shared the encouraging news that he was being introduced to new top executives of multinational corporations. "I am endeavoring to meet the treasurer of Esso of Indiana or Amoco... I am doing everything I can... [to be] a catalyst between businessmen and the academic world" (Stedman Jones, 2012: 135–6). These contacts led to the founding of the Manhattan Institute for Policy Research in 1977 and the Pacific Institute for Public Policy in 1979. In 1981 Fisher went on to found the Atlas Economic Research Foundation, the purpose of which was "to litter the world with free-market think-tanks." By 1991 Atlas had spawned at least 78 institutes around the world (Cockett, 1995: 307).

It is ironic that from Austria to the LSE, from the MPS and the IEA to the tight-knit web of scholars, wealthy businessmen, the media, and politicians, the entire neoliberal enterprise that originated from Hayek's distaste toward Nazism is flourishing in the 21th century under the command of an American family that helped Nazi Germany to build its war time oil refinery—an "investment" that not only made the family's fortune but helped to promote Nazi aggression (Mayer, 2016: 29–31). The way that democracy is being practiced in the 21th century cannot be appropriately understood without having a grasp of the "Kochtopus." The influence of Hayek on Charles Koch, a billionaire businessman and an extra-heavy-weight figure in the climate denial business, has been profound. In a speech given to the Institute of Humane Studies in 1997, entitled "Creating a Science of Liberty," Koch talked about his intellectual pilgrimage for a free society: "Probably the first book on liberty I read was Leonard Read's *Elements of Libertarian Leadership* in 1963." Read was the founder of the Foundation for Economic Education, an admirer of Hayek, and a core member of the MPS from its first meeting. Read's work inspired Koch not only to develop "passionate commitment to liberty," which he saw as "the form of social organization most in harmony with reality and man's nature," but also resulted in his being "exposed in-depth to thinkers such as Mises and Hayek." Von Mises and Hayek, stated Koch, "have had the greatest influence on my thinking.... Following Hayek's model of the free society as an experimental discovery process, I have engaged in a large variety of activities to advance the free society over the past 30 some years" (Koch, 1997). He then provided a long list of his endeavors emulating Hayek's model:

> In 1964, for example, Baldy Harper recruited me to help develop the Institute for Humane Studies[11].... In the same period, I began supporting a large number of market-oriented intellectuals such as F. A. Hayek... In those early years, I focused entirely on helping scholars develop because I believed knowledge was the key to progress.... I've supported so many hundreds of scholars with so many different approaches because, to me, this is an experimental process to find the best people and strategies.... In 1977, I recruited Ed Crane to convert the former Charles Koch Foundation to the Cato Institute in order to develop public policy tools from market concepts.... I provided the seed money for the Institute for Justice, and support for a wide range of other market-oriented organizations including.... American Legislative Exchange Council (ALEC)...., Pacific Research Institute...., Fraser Institute, and Heartland Institute. I have provided financial and educational support for free-market-oriented politicians.[12] (Koch, 1997)

A scene from the documentary film "Merchants of Doubt"[13] saw the CEO of the George C. Marshall Institute,[14] Bill O'Keefe, amused by the film director's question on what advice he would give to organizations such as Greenpeace on climate change, replying, succinctly, "they couldn't afford me!" Money is indeed crucial in building beliefs. It was made emphatically so under neoliberalism. In the battle of climate mitigation, however, the prominent role of money began not with the political contributions and PR fees paid by Big Oil to discredit climate science, but with Hayek's sympathetic business funders, who were impressed by his ideas about excluding sections of society while appearing to be enthusiastically including them. That seed money spawned neoliberal think tanks world-wide and successfully rewired public cognition about concepts of the individual, government, commons, equality, fairness, and liberty at the most profound level, long before climate change came on the scene. With superb skills in the art of exclusion through inclusion, neoliberals won the very expensive "competition of ideas," winning over the hearts of elites and ordinary citizens alike to promote and participate in the massive grabbing game while emptying democracy of its meaning and reducing it to the ritual of election. Not only was the upward redistribution of wealth implemented resistance-free, the accompanying destruction of nature also proceeded with the giddy consent of the public. Today, this modern-day enclosure of the commons has left ever fewer resources for current and future generations to survive on, with those who master the art of exclusion sitting on piles and piles of resources, relatively unthreatened by the countervailing forces of society and nature.

Notes

1 Early in his career, Hayek was susceptible to the idea of government playing a significant role in ensuring orderly economy and good society. The more embroiled his career became in the business grabbing game the more steadfastly against government intervention he became.

2 Cockett (1994: 36).

3 "Hayek's vision… implied a fundamental acceptance of substantive inequality" which helped to justify the neoliberal insistence that inequality is not only unavoidable but even desirable (Stedman Jones, 2012: 63).

4 Hayek insisted that everyone had equal access to the market. As such, inequality was fine because social mobility was healthily at work. "[F]or anyone who lost out, their own initiative would give them the opportunity to succeed through repeated attempts" (Stedman Jones, 2012: 63).

5 Orwell went on to complain that except for the few surviving commons, "every square inch of England is 'owned' by a few thousand families. These people are just about as useful as so many tapeworms" that had found out a way of "milk-ing the public while giving nothing in return" (Orwell, 1996 [1944]: 207–8).

6 In 1939 Keynes again tried to bring attention to this by stating that he was ever more convinced that those seventeenth- and eighteenth-century thinkers exer-cised "deep wisdom" in discovering and preaching a profound linkage between personal and political liberty on the one hand and the rights of private property

and private enterprise on the other. He urged his readers to remain alert about the way the lawyers of the eighteenth-century perniciously twisted the notion into "the sanctity of vested interests and large fortunes" and recognize the truth which lied behind (1939, 121). Kingsley Marshall, who was in dialogue with Keynes in a piece titled "Democracy and Efficiency" in *The New Statesman and Nation*, elaborated on Keynes' point by adding that, in the period before the French Revolution, the notion of private property right that was historically linked to liberty referred to the right of the peasant to own the fruits of his own labor and of the man who ran a small business and made profit out of. While the right of personal property was without any doubt inseparable from the conception of liberty, it was extraordinarily confusing and misleading to identify the right to own the fruits of one's own labor with the right of Mr. Rockefeller to own the labor and to control the conditions of thousands of lives. "Surely the monopoly ownership of our day is one of the great enemies of liberty" (Keynes, 1939: 121).

7 Quoted in Cockett, 1995: 78.

8 Cited in Cockett, 1995: 104.

9 According to Stedman Jones, "Heritage was the link between Hayek and Ronald Reagan." The briefings of the Foundation "were meant to provide a primer on a particular issue that a congressman could digest on the train home in an accessible form that fit into his briefcase" (Stedman Jones, 2012: 163).

10 Monetarist economist Allan Meltzer summed up the importance of Milton Friedman by stating that he and his wife Rose Friedman swam against the strong current where voters, which were "more equally distributed than income," would have demanded a more equal distribution of wealth than envisioned by neoliberalism. Meltzer pointed out that the Friedmans could not stop or reverse this strong current, "but they influenced far more than most the ways in which people and politicians think and act" (Meltzer, 2003: 204).

11 The IHS had close ties with Hayek since its founding (Stedman Jones, 2012: 159–60).

12 Amusingly, immediately following the last point, Koch stated: "I ensured that Koch Industries followed a philosophy of profiting by the economic, not the political, means" (Koch, 1997).

13 Based on the book *Merchants of Doubt*.

14 Posing as an independent science policy think tank, the George C. Marshall Institute specializes in attacking climate science (Oreskes and Conway, 2010: 244).

References

Appleton, Samuel. (2016), "The problem with 'embedded liberalism': the World Bank and the myth of Bretton Woods." Working Paper No. 11, October 2016, The Centre for Global Political Economy, University of Sussex.

Blake, Robert. (1976). A Changed Climate. In Robert Blake and John Patten (eds.), *The Conservative Opportunity*. London: MacMillan.

Blundell, John. (1987). "How to Move a Nation," *Reason Magazine*, February.

Cockett, Richard. (1995). *Thinking the Unthinkable—Think-Tanks and the Economic Counter-Revolution, 1931–1983*. London: Harper Collins Publishers.

Galbraith, John Kenneth. (1965). "Came the Revolution," *New York Times*, 16 May.

Hayek, A. Friedrich. (2007 [1944]). *The Road to Serfdom—Text and Documents, The Definitive Edition*. Edited by Bruce Caldwell. Chicago, IL: The University Chicago Press.

Hayek, A. Friedrich. (1947). Hayek's paper, Mont Pelerin Society Archive, Box 14.

Hayek, A. Friedrich. (1960 [1949]). "The Intellectuals and Socialism," *The University of Chicago Law Review*. 417–433. Reprinted in George B. de Huszar (ed.), *The Intellectuals: A Controversial Portrait*. Glencoe, IL: The Free Press, 1960. 371–385.

Hartwell, R. M. (1995). *A History of the Mont Pelerin Society*. Indianapolis, IN: Liberty Fund.

Keynes, John Maynard. (1923). *The Tract on Monetary Reform*. London: Macmillan and Co.

Keynes, John Maynard. (1926 [1924]). The End of Laissez-Faire. In: J. M. Keynes. *Essays in Persuasion*. London: Palgrave Macmillan (2010).

Keynes, John Maynard. (1939). "Democracy and Efficiency," dialogue between Kingsley Martin and J. M. Keynes. 17, *New Statesman and Nation* 121–123.

Keynes, John Maynard. (1944). Keynes to Hayek, 28 June.*Collected Works*, Vol. 27, 385–388.

Koch, Charles. (1997). "Creating a Science of Liberty." Speech given at the Institute for Humane Studies. 11 January.

Mayer, Jane. (2016). *Dark Money—The Hidden History of the Billionaires Behind the Rise of the Radical Right*. New York: Doubleday.

Meltzer, H. Allan. (2003). Choosing Freely. In M. Wynne, H. Rosenblum, and R. Formaini, (Eds.). *The Legacy of Milton and Rose Friedman's Free to Choose*. Dallas, TX: Federal Reserve Bank of Dallas.

Mont Pelerin Society. (1947). Statement of Aims, 8 April 2019. www.montpelerin. org/statement-of-aims.

Oreskes, Naomi and Erik M. Conway. (2010). *Merchants of Doubt—How a Handful of Scientists Obscured the Truth on Issues from Tobacco Smoke to Global Warming*. New York: Bloomsbury Press.

Orwell, George. (1996 [1944]). As I Please. In Orwell, Sonia and Ian Angus (eds.), *George Orwell—The Works*. Vol. 12. London: Secker & Warburg.

Orwell, George. (1944). "Book Review: the Road to Serfdom by F.A. Hayek/The Mirror of the Past by K. Zilliacus," *The Observer*, 9 April 2019.

Polanyi, Karl. (2001 [1944]). *The Great Transformation: the Political and Economic Origins of Our Time*. Boston, MA: Beacon Press.

Ruggie, John G. (1982), "International Regimes, Transactions and Change: Embedded Liberalism in the Post-war Economic Order," *International Organization*, 36(2).

Shaxson, Nicholas. (2012). *Treasure Islands: Tax Havens and the Men Who Stole the World*. London: Vintage.

Skidelsky, Robert. (1996). *Keynes*. Oxford: Oxford University Press.

Skidelsky, Robert. (2011). LSE event. Keynes vs. Hayek. https://www.youtube.com/watch?v=PLBOKq4On7k.

Stedman Jones, Daniel. (2012). *Masters of the Universe—Hayek, Friedman, and the Birth of Neoliberal Politics*. Princeton, NJ: Princeton University Press.

von Mises, Ludwig. (1944). *Bureaucracy*. New Haven, CT, Yale University Press.

3 Political Science

Because climate change is man-made, social sciences have a critical role to play in climate mitigation, by explaining collective human behavior and devising ways to appropriately modify it. Until social scientists begin to take this responsibility seriously,[1] no amount of further research findings on the part of natural sciences will make an iota of difference as far as mitigation is concerned. Within this context, we can argue that *politics* alone holds the key to the survival of human and many other species. As Stehr observed, "climate change should not be seen primarily as an environmental or economic problem, but as a question of political governability of modern societies" (2016).

Political science, *the* academic discipline responsible for the studying of politics,[2] however, has remained disturbingly tight-lipped about climate change. Mainstream political science has demonstrated little sense of urgency in discovering how we got into our current mess and what we need to do to climb out of it. In political science journals and books, the overall presence of climate change barely reflects the severity of the crisis. Whereas the topic does feature, from time to time, in journal articles and books in the field, it is clear that overall political scientists are mostly preoccupied with other aspects of politics. Among the 629 research articles published in the decade between 2007 and 2016 by *American Journal of Political Science*—the leading journal in the field—just two directly dealt with the topic of climate change. The journal with the second highest impact factor, *World Politics*, published 202 research articles in the period of which none dealt with climate change. For *Review of International Political Economy*, the number was five out of 414. For *Political Analysis*, none out of 274, and for *Annual Review of Political Science*, five out of 237. This under-representation of the topic of climate change in mainstream political science journals indicates that the discipline as a whole does not see itself as having a crucial role to play, or a responsibility to assume, in fighting the climate crisis. The articles that the leading journals in any discipline choose to publish conveys important messages about which topics are deemed important within the mainstream of the discipline. As young scholars and graduate students develop a sense of what is "hot" in the field from the leading journals and build their careers by following the cue, the failure of climate change to make its way into influential political science journals is particularly worrisome.

A similar sign of paralysis in the discipline is also evident in the development of the "Kochtopus," (the influence of Charles Koch, a US billionaire and philanthropist for free market causes), which was touched on in the previous chapter. The lack of attention paid to the corrosion of democracy by a discipline that specializes in democracy is embarrassing. In this chapter I demonstrate that the types of paralysis found in both cases are closely related and that political science was at first a passive victim of the neoliberal expansion, but subsequently an active facilitating force that helped to undermine the appropriate functioning of democracy.

In a world where truthful information in the public interest was not hidden from the public, voters of democratic societies would never agree to rules that allowed wealth to become ever more concentrated in the hands of an ever smaller group of people. Democracy was therefore an obstacle to the entrenchment and sustained operation of the neoliberal massive grabbing game. The removal of such an obstacle called for the invention of further science-like rules to support the supposedly self-regulating market. The tale of the self-regulating market needed to be promoted as trumping not just all of the other variables in the work of the economy, but also all the concerns and objections raised against it during the normal functioning of democracy. Although economists comfortable with social inequality and exclusion had succeeded in subordinating society to the market *in theory*, because the study of politics and democracy did not fall naturally within the reach of economists, the subordination *in practice* inevitably ran into inconvenient disciplinary boundaries. The champions of neoliberalism saw the problem very early on and understood that the neoliberalization of democracy could not take place without also neoliberalizing the discipline that studied democracy. As a result, neoliberal economists began to provide political scientists with unsolicited guidance to study politics the "right" way—rewarded when preoccupied with the "right" topics, asking the "right" questions, adopting the "right" approaches, and punished when doing the opposite. They skillfully nudged political scientists to whittle down the notion that government was established to serve society and not the market. Over time, the belief that only when the market was happy could society be happy became a common implicit assumption in political science studies. Much to the delight of neoliberal champions, political scientists dutifully stayed preoccupied with topics distant from big-picture issues—including climate change—which risked raising fundamental questions disruptive to the neoliberal order.

The project of neoliberalizing political science is best understood as a seamless extension of the "free" competition of ideas between Hayek and Keynes, as discussed in the previous chapter. In 1945 Ludwig von Mises went to the U.S., where he taught at New York University (NYU) until 1969. During his tenure at NYU, it was the William Volker Fund, not the university, that paid his salary (Stedman Jones, 2012: 50). Under the leadership of Harold Luhnow, the Fund functioned as a fierce defender of intervention-free self-regulating market mechanisms. It would seek out "individuals who published arguments that corresponded with their stated ideals. After performing careful analyses of prior publications, they would send unsolicited financial support as a source of encouragement for further writing and research." It also "supported a series of institutions that were designed to foster communication among academics and students who opposed economic planning" (Burgin, 2012: 101). In 1950 Hayek went to the University of Chicago, with salaries paid by the Fund. Luhnow was keen to see returns on the money he invested, and Hayek was only partly successful in resisting the funders' control over the substantive work of his project.[3] Hence, rather than being "mere pecuniary accessories to the rise of the Chicago School," Luhnow and the Volker officers were "hands-on players, determined and persistent in making every dollar count, supervising doctrine as well as organization" (Van Horn and Mirowski, 2009: 157). Under such strong-handed supervision by his funders, Hayek started the Free Market Study at the University of Chicago that would "produce an American version of *The Road to Serfdom*... materialized in Milton Friedman's *Capitalism and Freedom*, published in 1962" (Stedman Jones, 2012: 92).

The funders' close supervision and hands-on approach paid off. Immersing students in works produced by the Mont Pelerin Society (MPS) was bound to create a new generation of neoliberal scholars devoted to spreading neoliberal ideas. One such student was James M. Buchanan. A socialist, or in his own words, "allocationist" and "grossly naïve" in his thinking about political alternatives prior to enrolling in the University of Chicago in the winter of 1946 (Buchanan, 2009: 141), Buchanan was transformed by the program and became one of the most dedicated and relentless neoliberal scholars who would later cooperate with Charles Koch in transforming the way that democracy worked in the United States and beyond. In a 1949 article Buchanan compared two political foundations for erecting government finance: the *organismic* theory, which conceived of the state as an organic entity, and the *individualistic* theory, which treated the state as nothing more than the sum of its individual members acting in a collective capacity. The gist of the article was that the *organismic* approach would run into insurmountable difficulty, as the state, tasked with maximizing "some conceptually quantifiable magnitude," would be at a loss because there was no feasible way to determine what was to be maximized. Vague concepts such as "social utility" and "social welfare" were unhelpful, as they offered no concrete guidance to governmental fiscal authorities as to what exactly to maximize. This difficulty, in contrast, was absent in the *individualistic* approach wherein the individual replaced the state as the basic structural unit. The state, having no ends of its own, owed its origin to—and depended for its continuance upon—the desires of individuals to collectively having part of their wants fulfilled. State decisions, therefore, were nothing more than the collective decisions of its individual members. "The extent and range of public services are determined by the collective willingness of individuals to purchase them. Services will be extended as long as the aggregate benefits are held to exceed the costs" (Buchanan, 1949: 498–9).

What Buchanan tried to tackle was a serious problem that neoliberalism ran into in subordinating society to the market: the state function whose role was to protect society from the market. Even though generous funding from corporations had made the dissemination of neoliberal ideas easy at the abstract level, translating ideas that advocated abandoning the needy and protecting the wealthy into concrete policies involving tax codes and welfare spending was hard. Packaging the ditching of the needy with the nice "organismic vs. individualistic" scholarly jargon, Buchanan was rolling out an innocent-looking roadmap for the materialization of neoliberal ideas in the policy world.

Forgetting that he had recently criticized the organismic approach for its unresolvable problem of defining social utility or collective good, Buchanan argued that, in the individualistic approach, "if the *society desires*," a fiscal system designed to maintain the status quo distribution of incomes, fiscal structures, tax-burden and allocation of public expenditures could be outlined and installed to approximate the individualistic ideal (Buchanan, 1949: 504). In plain English, if wealth was concentrated in the hands of a few, the status quo *could* be guarded through the correct design of the tax and expenditure system, if society so desired. In the decades that followed, the intellectual descendant of Hayek devoted his career to turning the following statements from an oxymoron into a paradox: The majority of a society cannot possibly agree to rules harmful to itself. The society, however, *can* arrive at such rules *if it so desires*.

As a logical next step to calling for the replacement of the organismic approach with the individualistic approach, Buchanan called for voting to be replaced with market forces as a way of arriving at collective social choice. In a 1954 article entitled "Social Choice, Democracy, and Free Markets," Buchanan asserted that "[v]oting and the market, as decision-making mechanisms, have evolved from, and are based upon an acceptance of, the philosophy of individualism which presumes no social entity" (Buchanan, 1954a: 117). He warned against the danger of voting resulting in a restriction of minority rights and pointed to his discovery of the absence of such danger where the market was the ruling system. The impingement on minority rights could take place in voting when the majority is composed of stable coalitions, which repeatedly outvote the minority. Under such a tyranny of the majority, minorities might revolt against majority decisions. Using tax bills as an example, Buchanan asserted that from the individual minority member's point of view, a majority decision is not different from an authoritarian decision. In both cases, the decision was dictated to the individual, given that his values were overruled by the decision-making (Buchanan, 1954a: 120). The beauty of the market, in contrast, was that it did not require individuals to make a collective decision at all. Unlike ordinary political voting, which overruled minority values, market choices were achieved without impinging on individual values (Buchanan, 1954a: 122–3).

Buchanan's view about the merit of replacing democracy with market was an extension of von Mises' concept of the "sovereignty of consumers." In his 1949 book *Human Action*,[4] von Mises had already made a comparison between market and democracy:

> With every penny spent the consumers determine the direction of all production processes and the minutest details of the organization of all business activities. This state of affairs has been described by calling the market a democracy in which every penny gives a right to cast a ballot. It would be more correct to say that *a democratic constitution is a scheme to assign to the citizens in the conduct of government the same supremacy the market economy gives them in their capacity as consumers.* However, the comparison is imperfect. In the political democracy only the votes cast for the majority candidate or the majority plan are effective in shaping the course of affairs. The votes polled by the minority do not directly influence policies. *But on the market no vote is cast in vain.* Every penny spent has the power to work upon the production processes.... It is true, in the market the various consumers have not the same voting right. The rich cast more votes than the poorer citizens. But this *inequality is itself the outcome of a previous voting process.* To be rich, in a pure market economy, is the outcome of success in filling best the demands of the consumers. A wealthy man can preserve his wealth only by continuing to serve the consumers in the most efficient way (1949: 271–2, emphasis added).

Von Mises' idea of replacing "one person one vote" with "one dollar one vote" was merely an expression of his personal value, which was probably unconvincing or even perverse to most people. Buchanan, in contrast, was talented in turning value of this sort into something bearing strong resemblance with objective scholarly works. Buchanan argued that the market was superior to voting on multiple grounds: In the market, the individual could predict the result of his action with absolute certainty. The responsibility for a choice would remain undivided and uniquely concentrated on the chooser. Moreover, choices in the market were not mutually exclusive— that is to say, the selection of one product did not preclude the selection of another. Individual choice in the market could therefore be more articulate than in the voting booth. An extension of these features of the market meant the absence of coercion, which for Buchanan was the very definition of freedom. By the same token, "unfreedom" should be defined as the state of being prevented from utilizing the normally available capacities for action (Buchanan, 1954b: 340).

Posting as an open-minded and impartial scholar, Buchanan gave the pretense that he could accept voting as the superior form of individual choice in comparison with the market. This could be the case, however, "only if men [we]re able to agree on the ultimate social goals," (1954b: 341) which he clearly deemed impossible, given the desire of the wealthy few to preserve and further expand their wealth and the desire of the majority to achieve social equality. No such pretense was necessary when he was addressing his MPS fellows. His less-guarded paper presented at the 10th meeting of the MPS in 1959 gives us a fuller view of Buchanan's agenda. Buchanan reminded his fellow MPS economists that there was a fundamental blind spot in the approach of economists, which needed urgent attention. Economists, he argued, had overlooked the implications of the fact that public policies could emerge only from some collective choice process. It had been, and continued to be, a "gross error" to assume that after policy criteria were discussed and defined, policies would simply be formulated accordingly, independent of the collective decision-making process. Buchanan asserted that, owing to the abysmal ignorance of politicians as embodied in tariffs, restrictions on trade, controls over wages and prices, minimum wage laws, agricultural subsidies, and instability in the value of money, individual and market freedom continued to be constrained. Owing to such "deliberate machinations of men who, at least by our own scale of values, [we]re essentially 'evil'," public policy formation had, despite diligent works by his fellow MPS economists, not changed for the better (Buchanan, 1960: 265).

What this MPS paper reveals is that Buchanan's other journal articles, which came across as objective and unbiased, were in fact motivated by a desire to generate massive social, political, and economic changes that would better serve the interests of the wealthy few. As will be seen, this agenda would continue throughout Buchanan's academic career. Unlike normal scholarly inquiries, Buchanan's research agenda began not with the observation of a phenomenon needing investigation and explanation. Instead, it began with a firm conclusion mandated by business funders of the MPS and its satellite think thanks and institutions, including Buchanan's own Thomas Jefferson Center for Studies in Political Economy (later known as the Center for the Study of Public Choice) founded in 1957. It was from these firm conclusions that Buchanan worked his way back to theory-inventing and inquiry-formulation. Buchanan's powerful insight was that, if the roadblock to the success of business-friendly public policies was the "collective decision-making process," then what was needed was the invention of theories that could "rectify" this collective decision-making process. It was the rules for rulemaking, in other words, that needed to be placed in order. Efforts must be made to secure some changes in the political constitution: "The maintenance of the free society may depend upon the removal of certain important decisions from majority-vote determination." Hence, for issues involving major economic interest, decision-making must shift from majority rule toward greater consensus (Buchanan, 1960: 276). In plain language, democracy was a major enemy for those who excelled at grasping, because it stood in the way of establishing and maintaining rules that gravitated toward inherently exclusionary and predatory acquisition. To neoliberals, preventing those who were good at grabbing from unbridled grabbing was the most abhorrent form of unfairness. Unsurprisingly, sugarcoating the predatory and exclusionary neoliberal agenda into a magic "win-win," resistance-proof formula became the most challenging task for scholars such as Buchanan.

The public choice theory that Buchanan went on to create would have unfortunate, far-reaching effects on the way that democracy worked, wealth was distributed, and issues such as climate change were handled. The vision of Hayek and von Mises about how the world *should* work was for a long time popular only within a small circle. Even though the investment of sympathetic business funders helped to enlarge the circle, there was a limit to the extent to which people could support extremely business-friendly and blatantly biased public policies. Buchanan was able to break this limit by rewriting the rules of rule-making. Disguising his real agenda, Buchanan eventually transformed the way that the world *did* work through academic works, tailor-made for his and his funder's purpose, which came across as scientific, objective, and impartial.

In 1958 the William Volker Fund, the primary funder of James Buchanan's Center at the University of Virginia, sent him a postdoctoral fellow "with the small wrinkle that [Gordon Tullock] had no doctorate in economics or a plan to earn one" (MacLean, 2017: 76). *The Calculus of Consent: Logical Foundations of Constitutional Democracy,* co-authored by Buchanan with Tullock and published in 1962, was an essential playbook for rewriting public decision-making and constitutionalizing neoliberal order, first in America and later across the world. While Dwight Lee, in his article "The *Calculus of Consent* and the Constitution of Capitalism," which appeared in the Cato Institute's journal in 1987 (Lee, 1987: 332), hailed the book as "protecting capitalism from government," Nancy MacLean thought "[i]t might more aptly be depicted as protecting capitalism from democracy," the key message being that representative government "would destroy capitalism by fleecing the propertied class—unless constitutional reform ensured economic liberty, no matter what most voters wanted" (MacLean, 2017: 81).

The subtitle of the book—*Logical Foundations of Constitutional Democracy*—already revealed that the core issue that the authors intended to investigate was not economy but democracy. Contrasting *economic* theory with *political* theory, the authors asserted that the former, being a theory of collective choice, "provides us with an explanation of how separate individual interests are reconciled through the mechanism of trade or exchange," whereas the latter had up to that point failed to "[consider] fully the implications of individual differences for… political decisions" (Buchanan and Tullock, 1962: 3). Postulating that collective action must be composed of individual actions, these economics-turned-democracy experts took as the starting point of their theorizing "the acting or decision-making individual as he participates in the processes through which group choices are organized." Under this "methodological individualism" (1962: 3, 265), the first crucial question to ask is: "What do individuals want?" The obvious answer to them was that "the representative or the average individual, when confronted with real choice in exchange, will choose 'more' rather than 'less' " (1962: 18).

The failure to use this essentially economic approach, claimed the authors, explained the failure of political theorists to properly understand collective activity. "Their analyses of collective-choice processes have more often been grounded on the implicit assumption that the representative individual seeks not to maximize his own utility, but to find the 'public interest' or 'common good' " (1962: 20), when the plain truth was that "[a] shift of activity from the market sector cannot in itself change the nature of man" (1962: 306). Politics, they argued, should be seen as a form of "exchange." Given that it was the *same* individual—the utility maximizer—that participated in both the market process and the political process, the best prediction for the behavior of a representative, argued Buchanan and Tullock, was what was electorally rewarding rather than "selfless pursuit of the 'public interest' or the 'general welfare' as something independent of and apart from private economic interest" (1962: 283). From this perspective, politicians must be understood to be making decisions in accordance with the maximization of their own self-interests even while packaging such decisions with rhetoric of "common good" and "general welfare." This understanding of the process of collective choice brought into question the sacrosanctity of the majority rule in democracy. "Special interest" could easily pose as "public interest" and get its way by cajoling vote-seeking politicians. Given their implicit worry that the interests of the rich minority could be sacrificed, the authors stressed that the simple fact that "'public interest' is what the individual says it is" must be recognized, leading to the conclusion that only decisions with unanimous support could be defined as "in the public interest" (1962: 284). Buchanan and Tullock "hope especially," that their "theoretical construction will cause the student of political process... to consider more carefully and more cautiously the proper place of majority rule in the constitutional system" (1962: 301). Instead of being sacrosanct, majority rule should only be "one rule among a continuous set of possible rules for organizing collective decisions" (1962: 302). By "introducing some rule for unanimity or full consensus at the ultimate constitutional level of decision-making," "the institutional setting for collective choice-making [can] be constructed in order to confine the exploitation of man by man within acceptable limits" (1962: 6, 305).

After repeatedly hinting at and demonstrating the "right" way to conduct political science research through works such as *The Calculus of Consent*, Buchanan finally resorted to explicitly denigrating political science in 1966. Painting the extremely value-laden neoliberalized economics as value-free, objective, and scientific, Buchanan then presented political science as backward and inferior compared with economics. With political scientists as the targeted readers, Buchanan asserted that "the important theoretical advances in the explanation of political phenomena have been made primarily by those who approach the subject matter as economists." The reason that the political scientist failed to do his jobs properly was because he

> has not, traditionally, incorporated a theory of human behavior into his structure of political process. To him, "theory" has never implied prediction. Instead, political theory has suggested normative philosophical discourse on the objectives and aims of political order. Little, if any, positive science is to be found in this tradition. (1966: 176–7)

In contrast, the economist,

> shifting his attention to man's behavior in reaching collective decisions in concert with his fellows in some political arrangement, brings with him, ready-made so to speak, a basic behavioral postulate. He is able, through its use, to make predictions, to advance hypotheses that are conceptually refutable.... He can claim that he has a "theory of politics," of the way men do behave in collective decision making. (1966: 177)

Until the political scientist possesses as the economist does the underlying theory of human behavior—the belief that "individuals, when confronted with effective choice, will choose more rather than less," he remains like a "jibbering idiot" making "only noise under an illusion of speech" (1966: 170).

This shrewdly planned nurturing and indoctrination of selfishness in the way that the democratic political system was analyzed and accordingly designed was critical for the spreading of the massive grabbing game in the past half-century. The nearly insurmountable difficulty that we face today in altering planet-destroying behavior is fundamentally linked to the way that we study, understand, and practice democratic politics. This neoliberal way of studying, understanding, and practicing democracy was the intended consequence that neoliberal scholars such as Buchanan wanted when they set out to conquer neighboring disciplines through "economics imperialism".[5]

By taunting political science as unscientific in the 1960s, Buchanan pressed the right button at the right time. After World War II, political science as a discipline went into what Dahl called the "behavioral mood" (Dahl, 1961: 766; Easton, 1963 [1953]: 37). It was, according to Dahl, a protest movement that arose from "a strong sense of dissatisfaction with the achievements of conventional political science, particularly through historical, philosophical, and the descriptive-institutional approaches" (Dahl, 1961: 766). The fear of being alienated from the other social sciences such as economics and the aspiration to be "truly scientific" gave rise to this self-questioning. To rescue political science from the status of being "the least advanced" of all social sciences (Easton, 1963 [1953]: 40), Easton urged political scientists to adopt scientific methods and to strive for universal generalizations about social relations in order to "fulfill the promise in its designation as a political *science*" (1953: 5). For a method to qualify as scientific, theories must be built from variables that were "operationalizable," and hypotheses must be "testable" against empirical data (Ricci, 1984: 135–9). In addition, theories must be formulated in a way that they were "falsifiable," meaning that possibilities must exist for other existing cases to invalidate the theories (King, Keohane and Verba, 1994: 19).

All these newly invented yardsticks for measuring the authenticity of scientific research in political science were predicated on the dichotomy between the scientific study of what *is* and the nonscientific study of what *ought to be* (Ricci, 1984: 136–7).[6] The former was clear, non-controversial, and thus authoritative. The latter was subjective, messy, even authoritarian. In Easton's own words, behavioral research tended to be "explanatory rather than ethical" (Easton, 1962: 25). This dichotomy demonstrated a misunderstanding or disregard of the word "science" in the term "political science." For Aristotle, political science was different from *contemplative* science such as physics and *productive* science such as arts and rhetoric. Although contemplative science was concerned with truth or knowledge for its own sake and productive science with the making of useful or beautiful objects, *practical* science was concerned with "conduct and goodness in action, both individual and societal" (Shields, 2016). As politics was about "the noble action or happiness of the citizens" (Miller, 2017), the study of it should be normative and prescriptive rather than purely empirical and descriptive.

The results of misrepresenting politics as determined by some universal laws on a par with gravity and deeming the discovery of such laws as the sole job of political scientists are disastrous as seen from today's climate-change-stricken world. Although knowledge derived from contemplative science has an important role in politics, such knowledge cannot replace negotiation, persuasion, reasoning, and argument, which are essentially what politics is about. Contemplative science, in other words, has its limitations; it cannot define for a society what is just and socially desirable. The tragedy of political science in the past half-century is the failure to admit this plain and simple fact.

Rejecting Aristotle's understanding of political science, Easton argued that, just as knowledge of gravity can help us to build stronger bridges to enhance safety, and knowledge of atmospheric laws can help us to seed clouds to produce rain, in the social sciences, knowledge of generalizations can help us to "modify the conditions of external existence" (1963 [1953]: 30). Easton's ideal brand of political science amounts to focusing the analysis of politics on a realm emptied of chance, conjecture, emotion, and human variability: in other words, the political. There is no doubt that scientific knowledge *can* help us to take necessary actions, as Easton's examples clearly demonstrated. Yet a significant part of politics is about whether governments *will* take certain actions: Will the government collect enough tax to make bridges safer, even when they are located in low-income areas? Will the government invest in effective measures to reduce frequencies of extreme weather, even when such investment runs counter to the interests of big corporations? Where objective and scientific knowledge is available, the interface through which such knowledge is (or fails to be) transformed into public policy is profoundly political. Were crucial scientific knowledge automatically translated into public policies, climate change would not be threatening the survival of human civilization today, because works by scientists such as James Hansen and Michael Mann would have generated emission-curbing policies in the 1980s.

Stripping politics—the very subject political scientists were supposed to analyze—out of the discipline, the search by behavioralists for universal generalization inevitably ran into the problem of self-fulfilling and self-denying prophecy (Easton, 1963 [1953]: 24–9). Unlike natural sciences where the scientific discovery of universal laws such as the law of gravity cannot result in the law being altered, in social sciences, the discovery of supposedly universal laws governing human behavior can and often does lead to behavioral change, thus validating (self-fulfilling) or invalidating (self-denying) the claimed universal generalizable theory. Despite such irrefutable differences between natural and social sciences, behavioralists ignored criticisms that it was inadequate at a fundamental level to approximate research methods of social sciences to those of natural sciences: Humans *react* to scientific claims about their behaviour. The sun, the moon, the light, and the climate do not. While Easton cautioned that universal and enduring theories in social sciences should "*pattern after* but not *slavishly ap[e]* the laws of the natural sciences" (1963 [1953]: 31, emphasis added), he failed to elucidate on the difference between "patterning after" and "slavishly aping." Easton adopted a straight-line view of knowledge, which implied that natural sciences were ahead of social sciences and the latter could catch up by mimicking the research method of the former.

> I am not suggesting, of course, that it is either probable or possible that we can *immediately* chance upon a body of theory that even approaches in architectural form or intrinsic explanatory value the theories of such natural sciences as physics, chemistry, or biology. To pose as an *immediate* goal the attainment of the methodological rigor and precise formulations of the physical sciences, which are *centuries ahead* of the social sciences in their theoretical and factual maturity, would be to fall victim to scientism, the premature and slavish imitation of the physical sciences.[7] (Easton, 1963 [1953]: 59, emphasis added)

Despite the caveat, Easton clearly saw a need to intensify the process: "[A] civilization has seldom been faced with crisis weighted with graver consequences than that confronting us today.... The fund of political knowledge falls far short of what is required" (Easton, 1963 [1953]: 40). It is a great irony that these words ring more true today for reasons exactly opposite to what Easton had in mind. The rush to model social science research on that in natural science undermined the ability of society to see physical realities that are in fact scientific.

To understand this, it is important to point out that scientific claims are themselves an integral part of politics, and so are routinely pursued for goals other than truth alone. Just as the donning of a "lab coat" turned the scientist into a nonpartisan spokesperson of truth in Latour's story (Latour, 2004: 64), policies carried more weight if they were prescribed by political scientists claiming to have adopted scientific research methods. The implications of the presence or absence of scientific claims were profound for the relocation of resources. Policies that seemed to a layperson to be predatory and unjust could now be safely implemented once backed by a theorem or a law that had been reached "scientifically." The opinions and instincts of the non-scientific layperson can be deemed unsophisticated, backward, and lacking in scientific grounding. Such opinions must be kept at a distance from the sanctity of science. This volume-enhancing effect of the "lab coat" was appealing to those participating in public policy dialogues. In these terms, the self-fulfilling and self-denying prophecy problem was not a problem at all, but an opportunity for a researcher to instigate public behavior that he favored and dissuade public behavior from that which he disliked. Science, in other words, was rendered an instrument not for seeking truth but for gaining influence. By fostering a scientific repute for political science, avant-garde behavioralists had as their objective to mold the behavior of the public.

Someone who followed this logic was Harold Lasswell. Easton, himself an early behavioralist,[8] spoke about Lasswell's pioneering role in the transformation of the discipline (Easton, 1963 [1953]: 22, 75, 81). Dahl similarly took Lasswell's work as "prime examples of the behavioral approach" (Dahl, 1961: 767). Critics likewise named Lasswell one of the "founders of modern political science" (Chomsky, 1999: 55). Given Lasswell's influence in instigating the scientific revolution of political science, his central concern was commonly deemed to be with methodology. This misunderstanding "stems from the failure of his readers to relate its different parts to [a] central theme... for a careful consideration of the entire body of his writings can hardly fail to reveal the central significance of propaganda and the propagandist" (Horwitz, 1962: 242). Horwitz considered "social control through science" to be a decisively important aspect of Lasswell's work (1962: 237), and what propelled Lasswell to commit to the goal of social control was his paternalistic understanding of democracy. In his review of Walter Lippmann's *The Phantom Public*, Lasswell agreed with Lipmann's portrait of the public as "spasmodic, superficial, and ignorant," but thought that the book fell short of offering a practical solution. Noting that the public was marked by "active sentiments and conceptions about right and wrong," and a tendency to opinionate about policies, Lasswell saw Lippmann's hope in "exhorting the public to quit meddling when it feels an impulse to interfere" as wishful thinking. Instead of patiently exhorting and waiting for the ignorant public to turn quieter, "the most skillful or the most brazen publicity man, the best access to the most space in the newspapers... ought to out-brazen the rest" (Lasswell, 1926: 534–5).

In fulfilling their duty to outbrazen the rest, it was the instrument of propaganda that the wise few must depend on.

> [P]ropaganda will in time be viewed with fewer misgivings. At first sight its practise by specialists would appear to clash irreparably with some fundamental canons of a society which calls itself democratic. Such are the theory that the individual is obliged to participate openly and continually in ascertaining the general will and the theory that one who regardless of his private opinion propagates a view for a client commits a breach of obligation. The propagandist can show, however, that even a democratic society permits exceptions. (1934: 526)

In fact, "[t]he modern conception of social management is profoundly affected by the propagandist outlook... [which] combines respect for individuality with indifference to formal democracy" (1934: 526–7).[9]

> *This regard for men in the mass rests upon no democratic dogmatisms about men being the best judges of their own interests. The modern propagandist*, like the modern psychologist, *recognizes that men are often poor judges of their own interests*, flitting from one alternative to the next without solid reason. (1934: 527, emphasis added)[10]

These extensive quotations from Lasswell's support Horwitz's claim that Lasswell was chiefly concerned not with methodology, but with social control through science.[11] It is important to point out, however, that despite the views that he shared with Lippmann on the stupidity of the public, Lasswell was not a neoliberal. Horwitz used "radical egalitarianism" to describe the substance of Lasswell's politics (1962: 295), which he summarized as "a plea for greater equality, freedom from various forms of economic exploitation, and the minimization of social strife through intelligent social action designed to maximize the welfare of the many and reduce human suffering" (1962: 237).[12] Ironically, as a result of his success in fostering credibility of science in political science, Lasswell has

> Served unwittingly as a tool of certain modern political philosophers who were also propagandists of a stature and subtlety beyond even his active imagination. While inviting the assistance of like-minded academics who long to pull "the strings of Punch and Judy," the aspiring puppet master cavorts upon a stage built by others. The Master Propagandist is himself the victim of Propaganda. (Horwitz, 1962: 303–4)

The fact that it was a "radical egalitarian" that was treating scientific method as a propaganda device made economists' timely intervention an imperative from a neoliberal perspective: What if the newly found authority of scientific political science encouraged behavior counter to neoliberals' interests? Appropriate guidance was necessary to stave off the encouragement of individual behavior detrimental to the neoliberal ideal and to graft pro-grabbing rules for rule-making onto the discipline of political science.

Exploiting the self-doubts of political science and the discipline's aspiration to become truly scientific, Buchanan equated traditional political science research with backwardness. From Buchanan's perspective, economics had provided a model for other disciplines to imitate if they wanted to become advanced and scientific. As the economist would not allow what was social to interfere with his adherence to the strictly individualistic "central principle" (the principle that "individuals would choose more rather than less" when confronted with effective choice), he now had in his possession an underlying theory of human behavior. Because of this, "he qualifies as a scientist and his discipline as a science" (1966: 169–70). To stop being a "jibbering idiot," the political scientist had better follow the economist and use a basic behavioral postulate to understand man's behavior in reaching collective decisions.

Buchanan conceded that not all political scientists were jibbering idiots.[13] He welcomed the development that a few political scientists had taken their cue from economists and were extending the simple principle that "individuals would choose more rather than less" to political decision making in a non-market context (1966: 181). One such political scientist was William Riker, who gave the discipline of political science its own brand of rational choice theory. He claimed that traditional methods used in political science could produce "only wisdom and neither science nor knowledge. And while wisdom is certainly useful in the affairs of men, such a result is a failure to live up to the promise in the name of political science" (1962: viii). Riker was so embarrassed by the backwardness of political science that he wrote that "the behavioral sciences are *sciences* only by the kindly tolerance of university faculties who are willing to put up with our pretensions and ambitions in appropriating the name" (1962: 6). To rescue political science from the *cul-de-sac* in which the discipline found itself, because of its traditional methods of "history writing, the description of institutions, and legal analysis" (1962: viii), Riker invented positive political theory, or rational choice theory, in order for political science to elevate the discipline "above the level of wisdom literature" (1962: viii). "By positive, I mean the expression of descriptive rather than normative propositions" (Riker, 1959, cited in Amadae and Bueno de Mesquita, 1999: 276). To achieve the goal of genuinely scientific study, Riker took economics as the main source of inspiration. The discipline, he thought, had spurred genuine sciences of human behavior, with the market and price system as the proof of success for creating coherent theory and verified generalizations (1962: viii). The micro-foundations of *politics*, such as individual decision-making, were therefore considered the source of collective political outcomes and thus the appropriate starting point of research. By replicating methodological individualism and analyzing decision-making of rational (utility-maximizing) individuals, Riker believed that political scientists, too, could build models delineating generalizable relations between individual rational decisions and their collective political outcomes. Just as the trajectory of a particle could be predicted by knowing its momentum and the force affecting it, so the action of a rational decision-maker could be predicted by knowing the preferences of the actor and the environment structuring her choices. As a result, complex, strategic political interactions would become predictable, using models capable of projecting their stable or even law-like outcomes (Amadae and Bueno de Mesquita, 1999: 270–7). Mimicking the neoclassical theory of value in economics, which searched for market equilibrium sustained by balanced supply and demand, Riker compared voting to the market and advocated a general equilibrium outcome in politics (Aldrich, 2004: 323). Decades of research and a huge detour later, Riker arrived where political science had stood before the behavioral and rational choice revolution and declared that "[d]isequilibrium... is the characteristic feature of politics" (1980: 443). Having fundamentally transformed political science, rendering the discipline much worse equipped to face—or worse, an active facilitator of—challenges such as climate change, Riker now announced "politics is *the* dismal science," given that there were "no fundamental equilibria to predict" (1980: 443). "The power that comes from scientific knowledge is well nigh unobtainable for political abstractions," and "the wisest course for a science of politics is to abandon the search for equilibria" (1982: 200, 198). Speaking about traditional political science's emphasis on institutions with newly found respect, Riker invoked Aristotle and urged political scientists to re-emphasize the "classical heritage" of political science (1980: 433, 443).

In *How Culture Shapes the Debates of Climate Change*, Andrew Hoffman complained about academia being "a field of 'brick-makers'." In social sciences particularly, the fixation is rampant on generating lots of (often useless) bricks while remaining unconcerned about the cohesive whole that the bricks make up: "Today,... few social scientists are building an edifice, telling a whole story as it presently exists, and deciding what new pieces of information (bricks) may be necessary to tell the next chapter in the story" (Hoffman, 2015: vii). Within the political science community, its own members have long had the "nagging suspicion" that the discipline is merely "marking time" (Ricci, 1984: 212). It is hard to see how this would *not* be the case when the dominant approach has been methodological individualism relentlessly promoted by Buchanan, Riker, and their funders, which encourages fixation on the micro while undoing the *demos*.[14] Hoffman's "brick-maker" analogy came from Bernard K. Forscher's 1963 essay in *Science* magazine titled "Chaos in the Brickyard," in which he expressed concern that, over time, brick-making would become an end in itself (Forscher, 1963: 339). The regrettable state of scientific research that Forscher lamented in the early 1960s was the very template that pioneering behavioral and rational-choice theorists such as Easton and Riker used as the basis for the entire discipline of political science. While in 1963 Forscher worried about "an avalanche of random bricks," the production of which not only came "ahead of demand" but which also specified which kinds of bricks could "compete successfully with other brick-makers," in 1988 Riker celebrated political science's ability to look only at smaller events.[15] "Techniques without vision," a subtitle for the chapter on "Politics, Publishing, Truth, and Wisdom" in Ricci's *The Tragedy of Political Science—Politics, Scholarship, and Democracy* captivates the lamentable state of political science in the mid-1980s.

Pioneers in behavioral and rational choice theories were not unfamiliar with the criticism of triviality. Easton noted that critics have

> accused students of political behavior of selecting their problems not in the light of theoretical or ethical relevance but largely on ground of the accidental availability of technically adequate means for research. If a reliable technique is not at hand, the subject is not considered research-able. As a result... the argument runs, the behavioral approach is able to deliver reliable knowledge only with regard to political commonplaces or trivia. (1962: 11)

Easton's response to this powerful criticism was simply that it had misunder-stood behavioralism as the incorporation of scientific methods into political science, when in fact the behavioral turn was "basically a change in mood in favor of scientific methodology, methods, and techniques, with the emphasis on the latter" (1962: 11–2). This dismissive attitude toward a legitimate criti-cism partly explains why "exceedingly little" has been learned about the way that politics works after several decades of growth of rational choice theories, in terms of both complexity and sophistication (Green and Shapiro, 1994: ix–x). Green and Shapiro point out that, being interested mainly in theory ela-boration but not in empirical applications, proponents of rational choice tend to "[leave] for later, or others, the messy business of empirical testing" (1994: x). They point further to the "isolated and inward-looking" characteristics of rational choice theories, given that the enterprise thrives on "controversies of its practitioners' own making," rather than the real world societal problems that political scientists traditionally sought to understand (Green and Shapiro, 1994: x). "Obscurantism"[16] is Elster's term for describing the model-driven rather than world-driven works of theorists who treat models more as a "toy-box" than a "tool-box." It is important to note, as Elster has pointed out, that rational choice theory in general and game theory in particular can have great conceptual as well as practical value. The problem is the tendency that theor-ists, instead of aiming at truth, treat quantitative, mathematical, computer-based, or formal analyses as ends in themselves that are disconnected from their explanatory functions (2016: 2159–1).[17] Even after Riker admitted that the science of politics was after all very different from the science of economics and the analogy of election to the market was wrong, his obscurantism con-tinued elegantly, unapologetic about the social consequences of his advocacy and the obscure relationship between those consequences and the financial sources behind his advocacy:[18]

[I]f equilibria exist, prediction (and explanation) would also be relatively simple, thereby conforming to the scientific ideal of slicing the world up into self-contained pieces in order to predict and explain each piece by itself. The achievement of that ideal depends, however, on the pieces in fact being self-contained, and in the political case it depends specifically on the political process being independent of the morass of participants' perceptions and personalities, which are features of the world entirely outside the political abstraction. (1982: 199–200)

It is astonishing that it took several decades for a scholar like Riker to grapple with such a plain and simple fact. We can partially explain the indifference of the discipline towards the climate crisis through the influence of rational choice on political science, especially in terms of time marking, brick-making triviality, and obscurantism. However, this still does not begin to grasp the most profound meaning of the theory, which is directly linked to the neoliberal grabbing agenda. Rational choice theory is primarily a *normative* theory, offering advices to individuals about what to do, and only secondarily an explanatory theory (Elster, 2016: 2163). It was a powerful advocation of Buchanan's "individualistic approach" and a suppression of what he termed the "organismic approach" (Buchanan, 1949). Riker was credited first for leading political science *away* from its traditional method and then for leading it *back* to its traditional concerns, but the most significant "contribution" of Riker's great detour must be the destruction of the discipline's capacity (and indeed its will) to study, understand, and explain big-picture politics.[19] This explains its paralysis in discerning the corrosion of democracy, the rise of neoliberal expansion embodied in phenomena such as the "Kochtopus," and their interactive effects on the world's ability to handle climate change.

The relationship between political science and climate change is a quintessential example that removing politics from the central concern of the discipline in the name of science has resulted in the most anti-scientific and self-destructive collective behaviors in the part of the world where the development of rational choice political science has been the most advanced. The problem that some, or most, political scientists would find rational choice theory unpersuasive was easily taken care of. A number of brilliant political scientists might have made extraordinary contributions to big-picture climate solutions, had they not been preoccupied with producing "high impact-factor" journal articles. Indicators such as impact factor remain some of the most important neoliberal control tools for luring, nudging, herding, shaping, disciplining, purging, sifting, and excluding academics from direct influence. With such an effective disciplining system at work, rational choice theory did not need to make much sense for graduate students to subscribe to it, when career-wise, job prospects for those who did adopt the approach were clearly more promising. Amadae and de Mesquita marveled at the speed with which Riker's political science department at Rochester rose to become a top program of its kind: In 1959 the department of political science at Rochester did not graduate a single undergraduate major, but by June 1973 it graduated 26 doctoral students and 49 master's students. In the American Council of Education ratings, the department moved from being unrated in 1965 to the 14th in 1970. For the 1960–72 period, Rochester's student placement was second only to Yale's. These trailblazing first-generation Rochester PhD students "would be crucial in transforming the study of politics in the decades ahead" (Amadae and de Mesquita, 1999: 280).

How did a nascent neoliberal trailblazing program climb so fast? The secret of tending to the rules for rule-making, which Buchanan had long urged his fellow neoliberals to heed, was important. Accompanying the rise of rational choice theory was the rise of new "yard sticks" used for "measuring" the merits of one's research. Maske and Durden did a citation count of Riker's work and found that it has been cited "more than 3700 times by over 2000 different scholars in more than 500 different journals" (Maske and Durden, 2003: 191). For rational choice theory to be popular, it did not need to make sense, it needed to be cited, with the frequency expressed numerically.

This way of assessing the importance of one's work was not unique for academia. Rather, it was an effective neoliberal tool widely imposed on societies. As the "consent" that public services needed to prove their "competitiveness" in order to justify their existence was being manufactured by neoliberals with the help of corporate media,[20] governments started to "create simulacra of markets governed by economic or para-economic criteria of judgment in arenas previously governed by bureaucratic and social logics" (Rose, 1999: 146). By the late 1970s, the administrative doctrines that dominated public sector reform agendas began to converge on the importance of audit. Government programs started to problematize public-problem-solving and redirected emphasis on agencies meeting explicitly and concretely defined goals and targets, expressed in quantitative terms. The payment of government money to agencies would depend on quantitatively expressed audit results (Rose, 1999: 146).[21] In a chapter titled "audit explosion," Michael Power, author of *The Audit Society—Rituals of Verification*, noted that:

> During the late 1980s and early 1990s, the word 'audit' began to be used in Britain with growing frequency in a wide variety of contexts. In addition to the regulation of private company accounting by financial audit, practices of environmental audit, value for money audit, management audit, forensic audit, data audit, intellectual property audit, medical audit, teaching audit, and technology audit emerged and, to varying degrees, acquired a degree of institutional stability and acceptance... [A] growing population of 'auditees' began to experience a wave of formalized and detailed checking up on what they do. (1997: 3)

As Hood pointed out, the direct bearing of public choice theory on the shift from public administration to public management was clear and straightforward (Hood, 1991: 5). Just as the stated goal for promoting public or rational choice theory was a sharp deviation from its hidden goal, the stated goal of "better practices" to emphasize audit also served merely as a façade. Instead of encouraging "good" practices, audit trained auditees to produce "auditable" outputs. One of the most extreme examples is the "body count" in the Vietnam War that military commanders devised when Defense Secretary Robert S. McNamara demanded methods to quantify the war. In a 1977 survey of 173 American Generals who had served in the Vietnam War, 61% of the generals believed that the body counts were "grossly exaggerated" (Kempster, 1991).

As a system of "control of control" (Power, 1997: 82) and with the deliberately injected pressure of "competitiveness," "cost-cutting," and "doing more with less," audit helped to create the kind of collective mind-sets that conformed to the neoliberal vision. The alleged link between rampant audit and "accountability" was, therefore, illusive and highly problematic. Scott points out that, behind the rhetoric of transparency and impartiality, audit functions as a vast "anti-politics machine" that "[sweeps] vast realms of legitimate public debate out of the public sphere and into the arms of technical, administrative committees" (Scott, 2004: 119). However, far from eliminating politics, such practices "merely bury a vital politics in a series of conventions, measures, and assumptions that escape public scrutiny and dispute." Critically relevant to the current climate crisis is his observation that the audit practices

> stand in the way of potentially bracing and instructive debates about social policy, the meaning of intelligence, the selection of elites, the value of equity and diversity, and the purpose of economic growth and development. They are, in short, the means by which administrative elites attempt to convince a skeptical public—while excluding them from the debate. (Scott, 2004: 119)

Such techniques can therefore best be understood as "the hallmark of a neoliberal political order in which the methods of neoclassical economics, in the name of scientific calculation and objectivity, have replaced other forms of reasoning" (Scott, 2004: 119). This understanding easily explains Power's observation that the National Audit Office and the Audit Commission of the UK, both established by Prime Minister Margaret Thatcher, "became prominent forces in government, playing an evolving and complex *constitutional role*" (Power, 2000: 113, emphasis added).

With audit being an effective "control of control" playing a "constitutional role," placing education in general and higher education in particular under audit was only natural. An article celebrating Thatcher's legacy noted that "she felt the universities were complacent because they were over-protected from the market" and decided to introduce them to "greater accountability and... market forces." Concerned that "some universities were not using their research monies well," Thatcher introduced "accountability for research—a policy that became known as the Research Assessment Exercise" (Kealey, 2013). In time, professors, like other professionals, were required to "calculate their actions not in the esoteric languages of their own expertise but by translating them into costs and benefits that can be given an accounting value" (Rose, 1999: 152). As the "shift in the Zeitgeist towards the glorification of markets" (Karabel, 2005: 514, cited in Hazelkorn, 2017: 2) made ranking higher education a commercially profitable business, the "*US News and World Report* university ranking" emerged in the mid-1980s, further exacerbating the audit spiral. Soon, indices originally created for purposes *other than* audit and ranking became the holy grail for just about every young academic researcher. It is ironic that Eugene Garfield, who helped to create the Science Citation Index (SCI)—"the granddaddy of all citation indices" (Scott, 2004: 120), began his article "Citation Indexes for Science" published in *Science* in 1955 with this quotation:

> The uncritical citation of disputed data by a writer, whether it be deliberate or not, is a serious matter. Of course, *knowingly propagandizing unsubstantiated claims is particularly abhorrent*, but just as many naive students may be swayed by unfounded assertions presented by a writer who is unaware of the criticisms. (Williams, 1947)

It was for the purpose of eliminating the "uncritical citation of fraudulent, incomplete, or obsolete data" that Garfield thought of creating citation indexes for science. The way to eliminate such poor and irresponsible citation, he imagined, was through a citation index that would allow researchers to quickly check all works that had cited or criticized a particular paper. By bringing together works that would never have been collated by the traditional subject indexing, Garfield believed that the index that he was thinking about creating should best be described as "an association-of-ideas index." Through the Institute for Scientific Information (ISI) that Garfield founded in 1960, the SCI database was created in 1963, followed by the Social Sciences Citation Index (SSCI) and the Arts and Humanities Citation Index (AHCI). As a for-profit institute, the profitable ISI was sold to Thomson Scientific & Healthcare in 1992 and sold again to Clarivate Analytics in 2016.

Although the origin of citation index had little to do with neoliberalism, the neoliberal audit explosion that started to emerge in the 1970s quickly turned the citation index into something that it was originally created to eliminate. In "The Misuse of Numbers: Audits, Quantification, and the Obfuscation of Politics," Scott pictured a devoted university president employing the scientific techniques of quality evaluation in a truly comprehensive and transparent fashion. The scheme hinged on the citation indices: the SCI, the SSCI, and the AHCI.

> In keeping with the neoliberal emphasis on transparency, full public disclosure, and objectivity... the entire faculty is to be outfitted with

digitalized beanies… [The president] conjured a vision of the thrill students would experience as they listened, rapt, to the lecture of a brilliant and renowned professor whose beanie, while she lectured, was constantly humming, the total citations piling up before their very eyes. Meanwhile, in a nearby classroom, students worry as they contemplate the blank readout on the beanie of the embarrassed professor before them. How will their transcript look when the cumulative citation total of all the professors from whom they have taken courses is compared with the cumulative total of their competitors for graduate or professional school? (2004: 117)

With such a "single, indisputable standard of achievement" in place providing "a measure of quality and an unambiguous target for ambition," thought the president, "[j]unior faculty will no longer need to fear the caprice of their senior colleagues" (2004: 117–8).

This absurd picture only slightly exaggerated the reality of today's highly neoliberalized higher education. Although professors do not wear digitalized beanies that hum, they often denote which databases their published journal articles are indexed in and how high the "impact factors" of the journals are. Highlighting the "computer-like mindless" quality of such measuring scheme, Scott pointed out:

[s]elf-citations counted, adding autoeroticism to the normal narcissism that prevails in the academy. Negative citations, such as "X's article is the worst piece of research I have ever encountered" also count… Citations found in books, as opposed to articles, are not canvassed. More seriously, what if *no one ever read* the articles in which a work was cited, as was often the case? Then there is the provincialism of the exercise; this is, after all a massively English-language, hence Anglo-American, operation. (2004: 120)

In complaining how neoliberalized the discipline of political science has been "from methodology to epistemology to what we mean by a professional political scientist," Wendy Brown states that young graduate students are being taught in their first year—before they even have a thought about what political science actually studies—how to write publishable papers and how to entrepreneurialize their research, which "basically means repackaging in six ways for six different articles that are all the same but look like six different things on your resume" (Brown, 2015b).

The ability to tackle climate change is profoundly linked to the way that democracy is studied, understood, and practiced. The discipline most directly associated with the studying of democracy is political science, which informs public policies and public cognition alike in a fundamental way. The scientific revolution of the discipline has debilitated the discipline, as well as political systems informed by it in facing genuinely scientifically ascertained challenges such as climate change. No such clash between political science and natural science would be imaginable, had political science remained concerned with "conduct and goodness in action, both individual and societal" (Shields, 2016) and accepted the inevitability of being normative and prescriptive rather than purely empirical and descriptive. No allegedly universal laws or principles "scientifically" arrived at by political scientists would subordinate the relationship between carbon emissions and global warming, had political science stuck to Aristotle's understanding of politics as concerning "the noble action or happiness of the citizens" (Miller, 2017).

Lasswell's strategy of using science as an instrument, not to seek truth but to gain authority in propaganda, had a profound impact on the development of political science. In his presidential address to the American Political Science Association in 1956, "The Political Science of Science: An Inquiry into the Possible Reconciliation of Mastery and Freedom," Lasswell proposed to "inquire into the possible reconciliation of man's mastery over Nature with freedom, the overriding goal of policy in our body politic" (1956: 961). None of the concrete cases—from armament to evolution—that he provided, however, had begun to address the problem that Easton ran into when he used the examples of bridges and cloud seeding in justifying the scientific revolution of political science.[22] Although Lasswell hoped that, through science-certified propaganda, democracy could be protected from the stupid general public and remain within the control of enlightened elites like himself, he did not intend to use the science-certified propaganda machine to serve a neoliberal upward re-distribution system. The same cannot be said about the neoliberal rational choice strand.

In order to create a theory with a self-fulfilling effect helpful for forging a society where the selfish were rewarded while selfless altruists were punished, Riker framed the concept of rationality tactfully. His concept of rationality amounted to intentionally treating members of a society who failed to behave selfishly, whom he acknowledged as constituting the majority, as irrelevant. Noting that "[i]t must not be asserted that all behavior is rational but rather merely that some behavior is and that this possibly small amount is crucial for the construction and operation of economic and political institutions" (1962: 20), Riker went on to state:

> In the dynamics of pricing in a competitive market, there may be (and often are) sellers who offer at prices markedly different from that on which the market finally settles. Those who sell at less than the stabilized market price or who do not sell at all (and who in both instances may *fail to maximize like economic men*) may well constitute the vast majority of the sellers. But *it is not their behavior that counts*; rather it is the action of those sellers who do manage to offer at what turns out to be the stabilized price that, from the supply side, determine the nature of the market. If the market is thus controlled by those persons (possibly a minority) who behave in a maximizing way, then it can be said that the market institution selects and emphasizes rational behaviour. (1962: 20–21, emphasis added)

So, I believe, do other institutions, such as election systems, warfare, and other decision-making processes in which *several persons must, for the sake of winning, come together for common action without much regard for considerations of ideology or previous friendship.* Politics, in the old saw, makes strange bedfellows—and the very strangeness is the triumph of the maximizing motive in some (though far from all) participants. Since elections, warfare, etc., are decision-processes in which the stronger side wins, they place a premium on a side becoming stronger by any means possible which does not too flagrantly violate accepted cannon of behavior.... At any rate, these institutions that work by coalitions select and reward with success behavior which is apparently motivated by the intention to maximize power. (1962: 21, emphasis added)

The crude message underpinning this tactful framing of issues was that the world was not meant for non-grabbers. Moreover, those who did not wish to be selfish had no right to hold back those who did and who excelled at it. With this Hayekian exclusionary outlook, Riker, the person responsible for the immersion of hundreds of thousands of students in intensive rational-choice training and subsequently disseminating the individualistic and neoliberal way of understanding the world, was leading the discipline down a profoundly anti-democratic road. In his 1982 book *Liberalism Against Populism—A Confrontation Between the Theory of Democracy and the Theory of Social Choice*, Riker continued the Buchanan-style neoliberal attack on majoritarian democracy, arguing that the populism of Rousseau was untenable, and the fact that democratic voting and discussion were manipulable rendered democracy inaccurate and meaningless. As a result:

[t]he kind of democracy that thus survives [rigorous political science scrutiny] is not.... popular rule, but rather an intermittent, sometimes random, even perverse, popular veto. Social choice theory forces us to recognize that the people cannot rule as a corporate body in the way that populists suppose. Instead, officials rule, and they do not represent some indefinable popular will. Hence they can easily be tyrants, either in their own names or in the name of some putative imaginary majority. (Riker, 1982: 244)

Gerry Mackie examined the cases that Riker claimed were empirical evidence supporting his argument, "only to find that each is mistaken."[23] Mackie concluded that "that almost all the claims to demonstrate [the inaccuracy and meaninglessness of majority democracy] came from the pen of one man should have served as a warning signal. The anti-democratic interpretations of social choice results inspired by Riker, widely endorsed in the discipline of political science, are unsupported by evidence and must be abandoned" (Mackie, 2001: 23–4).

For over half a century, many so-called intellectual dialogues within the discipline of political science informed by rational choice theory had been engagements with public relations masterpieces that mastered the art of exclusion through inclusion, instead of genuine and honest scholarly works. As has been demonstrated in this chapter, the grand wild goose-chase itself was launched with a deliberate purpose. In the process of brick-making dialogues, the original core concerns of political science were successfully liquidated. Notions such as "common purposes" and "social justice" were tacitly purged from political science research. The discipline quietly transformed into an instrument serving neoliberal goals and came to play an active role in entrenching the neoliberal order in democratic societies, including the stalling of mitigation efforts in these societies. Man-made climate change, once scientifically proven as fact, would have had much less stamina without the neoliberal-fostered science revolution, aimed at purging notions such as "common purposes" and "social justice" out of the discipline. Imagine a world where political scientists were not trained to be foot soldiers for advancing the neoliberal art of exclusion through inclusion; imagine how it could have contributed to the prevention of climate tragedies through genuine, honest, and meaningful research on politics and democracy. Elinor Ostrom's *Governing the Commons*[24] has been used as an example for asserting that public or rational choice theory "works" and can thus help to "provide solutions" for the governing of the commons. However, the most notable feature of Ostrom's work was in fact how radically it, rational-choice-model ridden notwithstanding, *modified* the original core assumption of rational choice theory. Instead of accepting the rational choice trademark assumption that individuals always behave in the same way as economic actors do in the market, Ostrom made the harsh criticism that rational choice theorists' mistaking theoretical assumptions with reality made their theorizing *dangerous.*[25] It was by shaking the very ground that rational choice theory stood on that she seemingly saved the theory. However, her hope to "shatter the convictions of many policy analysts that the *only* way to solve common property resources problems is for external authorities to impose full private rights or centralized regulation" (1990: 182) was shattered when Buchanan published *Property as A Guarantor of Liberty*, trashing the idea of setting limits on the privatization of the commons. Buchanan argued that by partitioning shared resources in specifically assigned parcels and replacing common usage by private property, rational individuals would stop over-using but begin to optimize the usage of resources, thus ending "the tragedy of the commons" (1993: 6). While this pursuit of *economic efficiency* was an important element in justifying privatization, Buchanan claimed the protection of *individual liberty* to be an even more fundamental reason for partitioning and privatizing the commons (1993: 1–2). As a result, constitutional limits constraining political intrusions into voluntary contractual arrangements involving transfer of property must be put in place "prior to and separately from any exercise of democratic governance" (1993: 59).

Keynes was prescient when he concluded *The General Theory of Employment, Interest and Money* by saying:

> the ideas of economists and political philosophers, both when they are right and when they are wrong, are more powerful than is commonly understood. Indeed the world is ruled by little else. Practical men, who believe themselves to be quite exempt from any intellectual influences, are usually the slaves of some defunct economist. Madmen in authority, who hear voices in the air, are distilling their frenzy from some academic scribbler of a few years back. I am sure that the power of vested interests is vastly exaggerated compared with the gradual encroachment of ideas. Not, indeed, immediately, but after a certain interval; for in the field of economic and political philosophy there are not many who are influenced by new theories after they are twenty-five or thirty years of age, so that the ideas which civil servants and politicians and even agitators apply to current events are not likely to be the newest. But, soon or late, it is ideas, not vested interests, which are dangerous for good or evil. (1936: 167)

It is on the shoulders of neoliberal giants such as Hayek, Buchanan, and Riker that successful climate deniers from Fred Singer to Donald Trump stand.

Notes

1 Stehr, for instance, is surprised that "scientists' disenchantment with democracy and the implication that political liberties might need to be suppressed in light of profound future environmental changes has not received much systematic attention in social science" (Stehr, 2016).
2 According to the *Oxford Dictionary*, political science is "the branch of knowledge that deals with the state and systems of government," and "the scientific analysis of political activity and behavior."
3 Hayek was convinced that "the person who holds the purse strings will exercise a certain influence," and that one should not allow other concerns to lead one to abandon any essential source of support (Burgin, 2012: 99). See also Stedman Jones, 2012: 77.
4 The original argument was made in German in 1940. The English version was published in 1949.
5 Mäki pointed out, "'imperialism' in the case of economics imperialism... has been proudly adopted by the imperialists themselves with the purpose of celebrating it" (Mäki, 2009: 352). In a book chapter titled "Economic Imperialism," Gordon Tullock defined "economics imperialism" as "an attempt on the part of economics to take over all the other social sciences" and proudly announced the invasion of economists in neighboring disciplines (Tullock, 1972: 317). George Stigler, a key figure in the Chicago School in economics, took economics to be an "imperial science" that had been "aggressive in addressing central problems in a considerable number of neighboring social disciplines and without any invitations" (Stigler, 1984: 311). On economics imperialism, see Nik-Khah and Van Horn's article, which has a reprint of a poster for George Stigler's lecture in 1984 that featured an octopus with tentacles reaching towards other academic disciplines (Nik-Khah and Van Horn, 2012).

6   One of the key figures in the scientific revolution of political science, Harold Lasswell, distinguished between "the science of politics" and "the philosophy of politics." In his influential *Politics—Who Gets What, When and How*, Lasswell began the book by asserting "[t]he science of politics states conditions; the philosophy of politics justifies preferences. This book, restricted to political analysis, declares no preferences. It sates conditions" (1950 [1936]: 3). This approach was justified since "[t]he professors of moral philosophy were often so far out of touch with the new routines of business and government that they were at a loss for timely and relevant illustrations of classical aphorisms in terms of modern life" (1950 [1936]: 147). Given the discrepancy, new generations of specialists were keen to concentrate on "describing the new routines of the expanding civilization in which they lived. This greatly increased the academic emphasis upon the naturalistic at the expense of the normative" (1950 [1936]: 148).

7   Easton cited Hayek who defined "scientism" as "an attitude which is decidedly unscientific in the true sense of the word, since it involves a mechanical and uncritical application of habits of thought to fields different from those in which they have been formed. The scientistic as distinguished from the scientific view is not an unprejudiced but a very prejudiced approach which, before it has considered its subject, claims to know what is the most appropriate way of investigating it" (Hayek, 1942: 269).

8   Recall that the behavioral turn in political science arose owing to the dissatisfaction with conventional political research that took the historical, philosophical, and the descriptive-institutional approaches. Behavioral research sought to outgrow conventional political research by being explanatory rather than normative.

9   Lohrey pointed out that many imprudent academics, including Lasswell, had assumed the need to take the risk out of democracy by managing public opinion, "a management which has been in the interests of business." This academic and media conformity helped to close the American mind to "the kind of critical thought needed for a healthy, culturally diverse and pluralistic democracy. That this management of democracy should seem necessary and go unchallenged for so long in what is often hailed as the leading democracy in the world is a situation which reflects on the intellectual character of political and academic leaders in the United States" (Lohrey, 1995: 2).

10   The notion that the general public "must not intrude in the management of public affairs" and "keep to their function as interested spectators of action" for their own good (Chomsky, 1995: xi) would become a recurrent theme advocated by prominent political scientists. Samuel P. Huntington, for instance, "told a group of about 20 students... that a movement to attack established institutions in the 1960s had resulted in what he termed an 'excess of democracy,' and contributed to the breakdown of democracy in America..... Huntington said there was a worldwide pattern towards a general extension of democracy, which had resulted in non-democratic ends" (Dalton, 1976).

11   Citing Lasswell's first published work where he argued that "[o]ne aspect of the task of the systematic student of politics is to *describe* political behavior in those social situations which recur with sufficient frequency to make *prediction* useful as a preliminary to *control*," Horwitz summarized the intention of Lasswell's entire work as "description, prediction, [and] control" (1962: 227). Of the three "elements in this progression, the objective of social control is ultimate and governing... Scientific description is required for prediction, but prediction, in turn, is necessary for effective and intelligent social control" (Horwitz, 1962: 229).

12   Hence, unlike the rational-choice turn of the 1960s, the behavioral turn of the 1950s within political science should not be seen as a direct result of deliberate neoliberal manipulation.

13 By the same token, not all economists qualified as scientists. Keynes' modern macro-economic theory, for instance, was "really no theory at all," because by taking society as an organized community with common purposes capable of promoting economic and social justice, it had "divorced itself from the central proposition relating to human behavior" (1966: 170).

14 *"Undoing the Demos—Neoliberalism's Stealth Revolution"* is the title of Wendy Brown's book. "The Greek etymology of "democracy" .... *demos/kratia* translates as "people rule" or "rule by the people" (Brown, 2015a: 19).

15 The following quotation reveals the extent to which the cart of generalization had been placed before the horse of reality political scientists are supposed to explain in Riker's rational choice theory. "The difficulty with big events as units of analysis is that it is impossible to find enough of them out of which to construct precise analogies." This difficulty would result in "fuzzy generalization" with untestable assertion, no account of cause, and difficulty to tell just what features of the events one was generalizing about (Riker, 1988: 256).

16 Elster noted that there is "a less polite word for obscurantism: *bullshit*. Within Anglo-American philosophy there is in fact a minor sub-discipline that one might call *bullshittology*" (2012: 159).

17 Elster used Riker and Ordeshook's 1968 article, "A Theory of the Calculus of Voting" as an example. Riker and Ordeshook claimed to have presented empirical evidence that "citizens actually behave *as if* they employed" the calculus of voting they described in the article (1968: 25). To this habitual bold claim of rational choice theorists, Elster raises the banal question: *"[h]ow can one impute to real-life agents the capacity to make in real time the calculations that occupy many pages of mathematical appendixes in the leading journals and that can be acquired only through years of professional training?"* (2016: 2166, emphasis original) Even if rational actors can be assumed to have good learning abilities, absurdly large amounts of time would still have to be spent to ensure that a decision is optimal, which then become a case of hyperrationality and therefore of irrationality due to the high costs of decision-making itself (2016: 2168). Even on the simpler question of when a rational actor should stop searching for further information, rational choice theory has no good advice to offer. Elster points out that when rational belief formation is indeterminate, "one does indeed have to rely on intuition," a fact that renders the key assumption of rational choice problematic (2016: 2164–5).

18 Riker was among the active members of the Public Choice Society in its early years along with James Buchanan and Gordon Tullock. The society helped to generate the critical mass in establishing the rational choice approach as an "academy-wide method of inquiry" (Amadae and Bueno de Mesquita, 1999: 278). In the 1960s, the University of Rochester, where Riker built the Rochester school of political science, was "flush with capital" provided by head trustee of the Haloid-Xerox Corporation, Joseph Wilson, making the endowment second only to Harvard's and Yale's (Amadae and Bueno de Mesquita, 1999: 279).

19 "The antidemocratic interpretations of social choice results inspired by Riker, widely endorsed in the discipline of political science, are unsupported by evidence and must be abandoned" (Mackie, 2001: 24).

20 *Manufactured Consent—The Political Economy of the Mass Media* is the title of a 1988 book by Edward S. Herman and Noam Chomsky.

21 A recent book tackled the same problem. In *The Tyranny of Metrics*, Muller uncovers how the obsession with metrics, which encourages "teaching to the test" or "gaming the stats" is threatening the quality of our lives and important institutions (Muller, 2018).

22 See p. 42.

23   Mackie noted that "[i]n order to be persuaded to abandon the concept of the public good and the idea of democracy as in some sense the expression of the people's will, most people would require that it be robustly demonstrated that manipulation of outcomes be frequent, harmful, and irremediable." Riker's position is that it is either theoretically impossible or empirically difficult to detect such manipulation. He is able, however, to produce spectacular anecdotes that show harmful manipulation on major issues.

24   Through empirical studies on how small communities devised ways to care for shared natural resources such as pastures, fishing waters, and forests, Ostrom demonstrated the commons could be governed effectively without privatization or centralized state regulation. This book, published in 1990, won her the Nobel Prize in economic science in 2009.

25   "What makes these models so dangerous... is that the constraints that are assumed to be fixed for the purpose of analysis are taken on faith as being fixed in empirical settings" (1990: 6).

References

Aldrich, John. (2004). William H. Riker. Rowley, Charles and Friedrich Schneider (eds.), *The Encyclopedia of Public Choice,* Springer: New York, 321–324.

Amadae, S.M. and Bruce Bueno de Mesquita. (1999). "The Rochester School: The Origins of Positive Political Theory," in *Annual Review of Political Science.* Vol. 2, 269–295.

Brown, Wendy. (2015a). *Undoing the Demos: Neoliberalism's Stealth Revolution.* New York: Zone Books.

Brown, Wendy. (2015b). "Cultures of Capital Enhancement." Lecture at the Rev. Michael C. McFarland, S.J. Center for Religion, Ethics and Culture and the Center for Interdisciplinary Studies at Holy Cross. 12 November.

Buchanan, James. (1949). "The Pure Theory of Government Finance: A Suggested Approach," in *Journal of Political Economy.* Vol. 57, No. 6, 496–505.

Buchanan, James. (1954a). "Social Choice, Democracy, and Free Markets," in *Journal of Political Economy.* Vol. 62, No. 2, 114–123.

Buchanan, James. (1954b). "Individual Choice in Voting and The Market," *Journal of Political Economy.* Vol. 62, No. 4, 334–343.

Buchanan, James. (1960). "Economic Policy, Free Institutions, and Democratic Process," Paper presented at the 10th Meeting of the Mont Pelerin Society held at Oxford, September 1959. In *Politico.* Vol. 25, no. 2. 265–277.

Buchanan, James. (1966). Economics and its Scientific Neighbors. In Sherman Roy Krupp (ed.), *The Structure of Economic Science—essays on methodology,* Englewood Cliffs, NJ: Prentice-Hall, Inc., 166–184.

Buchanan, James. (1993). *Property as A Guarantor of Liberty.* Brookfield, VT: Edward Elgar Publishing Company.

Buchanan, James. (2009). Born-Again Economist. In Breit, William and Barry T. Hirsch (eds.) *Lives of the Laureates: Twenty-Three Nobel Economists.* Cambridge, MA: MIT Press.

Buchanan, James and Gordon Tullock. (1962). *The Calculus of Consent: Logical Foundations of Constitutional Democracy.* Indianapolis, IN: Liberty Fund.

Burgin, Angus. (2012). *The Great Persuasion.* Cambridge, MA: Harvard University Press.

Chomsky, Noam. (1995). Foreword. *Taking the Risk Out of Democracy—Corporate Propaganda versus Freedom and Liberty.* Edited by Andrew Lohrey. Urbana, IL: University of Illinois Press.

Chomsky, Noam. (1999). *Profit Over People—Neoliberalism and Global Order.* New York: Seven Stories Press.

Dahl, Robert. (1961). "The Behavioral Approach in Political Science: Epitaph for a Monument to a Successful Protest," *American Political Science Review.* 55(4): 763–772.

Dalton, Joseph. (1976). "Huntington Warns Breakdown due to Excessive Democracy," *The Crimson.* 24 March.

Easton, David. (1962). Introduction: The Current Meaning of "Behavioralism" in Political Science. In James C. Charlesworth (ed.), *The Limits of Behavioralism in Political Science.* Philadelphia, PA: The American Academy of Political and Social Science.

Easton, David. (1963 [1953]). *The Political System—An Inquiry into the State of Political Science.* New York: Alfred A. Knopf.

Elster, Jon. (2012). "Hard and Soft Obscurantism in Humanities and Social Sciences," *Diogenes.* 58(1–2), 159–170.

Elster, Jon. (2016). "Tool-box or Toy-box? Hard Obscurantism in Economic Modeling," *Synthese.* 193: 2159–2184.

Forscher, Bernard. (1963). "Chaos in the brickyard," *Science.* Vol. 142, issue 3590: 339.

Garfield, Eugene. (1955). "Citation Indexes for Science—A New Dimension in Documentation through Association of Ideas," *Science.* Vol. 122, 108–111.

Green, Donald P. and Ian Shapiro. (1994). *Pathologies of Rational Choice Theory—A Critique of Applications in Political Science.* New Haven, CT: Yale University Press.

Hayek, F. A. von. (1942). "Scientism and the Study of Society," *Economica.* 9(35), 267–291.

Hazelkorn, Ellen. (2017). Introduction: The Geopolitics of Rankings. In Ellen Hazelkorn (ed.), *Global Rankings and The Geopolitics of Higher Education— Understanding the influence and Impact of Rankings on Higher Education, Policy and Society,* 1–20.

Hoffman, Andrew. (2015). *How Culture Shapes the Climate Change Debate.* Stanford, CA: Stanford University Press.

Hood, Christopher. (1991). "A Public Management for All Seasons?" *Public Administration.* Vol. 69. 3–19.

Horwitz, Robert. (1962). Scientific Propaganda. In Herbert J. Storing (ed.), *Essays on the Scientific Study of Politics.* New York: Holt, Rinehart and Winston, Inc., 227–304.

Karabel, Jerome. (2005). *The Chosen—The Hidden History of Admission and Exclusion at Harvard, Yale and Princeton.* Boston and New York: Houghton Mifflin Co.

Keley, Terence. (2013). "How Margret Thatcher Transformed our Universities," *The Telegraph.* 8 April.

Kempster, Norman. (1991). "In This War, Body Count is Ruled Out," *Los Angeles Times.* 31 January.

Keynes, John Maynard. (1936). *The General Theory of Employment, Interest and Money.* London: Palgrave Macmillan.

King, Gary, Robert O. Keohane, Sidney Verba. (1994). *Designing Social Inquiry—Scientific Inference in Qualitative Research*. Princeton, NJ: Princeton University Press.

Lasswell, D. Harold. (1926). "Book Review: The Phantom Public by Walter Lippmann," *The American Journal of Sociology*. 31(4), 533–535.

Lasswell, D. Harold. (1934). Propaganda. In *Encyclopaedia of the Social Sciences*. Vol. 12. New York: The Macmillan Company, 521–527.

Lasswell, D. Harold. (1950 [1936]). *Politics—Who Gets What, When, How*. New York: Peter Smith.

Lasswell, D. Harold. (1956). "The Political Science of Science: An Inquiry into the Possible Reconciliation of Mastery and Freedom," *American Political Science Review*, 50(4), 961–979.

Latour, Bruno. (2004). *Politics of Nature—How to Bring the Sciences into Democracy*. Cambridge, MA: Harvard University Press.

Lee, Dwight R. (1987). "The Calculus of Consent and the Constitution of Capitalism," *Cato Journal*. Vol. 7, no. 2. 331–336.

Lippermann, Walter. (1922). *Public Opinion*. New York: Harcourt, Brace and Company.

Lippermann, Walter. (1925). *The Phantom Public*. New York: Harcourt, Brace and Company.

Lohrey, Andrew. (1995). Introduction. In Alex Carey, *Taking the Risk Out of Democracy—Corporate Propaganda versus Freedom and Liberty*. Edited by Andrew Lohrey. Urbana, IL: University of Illinois Press.

Mackie, Gerry. (2001). Is Democracy Impossible? Riker's Mistaken Accounts of Antebellum Politics. Manuscript.

MacLean, Nancy. (2017). *Democracy in Chains: The Deep History of the Radical Right's Stealth Plan for America*. New York: Penguin Publishing Group.

Mäki, Uskali. (2009). "Economics Imperialism: Concepts and Constraints," *Philosophy of the Social Sciences*. Vol. 39, no. 3, 351–380.

Maske, Kellie and Garey Durden. (2003). "The Contributions and Impact of Professor William H. Riker," *Public Choice*, 117:191–220.

Miller, Fred. (2017). Aristotle's Political Theory. In *The Stanford Encyclopedia of Philosophy* (Winter2017 Edition), Edward N. Zalta (ed.), https://plato.stanford.edu/archives/win2017/entries/aristotle-politics.

Muller, Jerry Z. (2018). *The Tyranny of Metrics*. Princeton, NJ: Princeton University Press.

Nik-Khah, Edward and Robert Van Horn. (2012). "Inland Empire: Economics Imperialism as an Imperative of Chicago Neoliberalism," *Journal of Economic Methodology*. Vol. 19, No. 3, 259–282.

Ostrom, Elinor. (1990). *Governing the Commons—The Evolution of Institutions for Collective Action*. Cambridge, UK: Cambridge University Press.

Power, Michael. (1997). *The Audit Society—Rituals of Verification*. Oxford: Oxford University Press.

Power, Michael. (2000). "The Audit Society—Second Thoughts" *International Journal of Auditing* Vol. 4, No. 1, 111–119.

Ricci, David. (1984). *The Tragedy of Political Science—Politics, Scholarship, and Democracy*. New Haven, CT: Yale University Press.

Riker, William H. (1959). Application to Center for Advanced Study in the Behavioral Sciences, Stanford University, CASBS file, William H. Riker papers, University of Rochester. Cited in Amadae and Bueno de Mesquita (1999: 276).

Riker, William H. (1962). *A Theory of Political Coalitions*. New Haven, CT: Yale University Press.

Riker, William H. (1980). "Implications from the Disequilibrium of Majority Rule for the Study of Institutions," *American Political Science Review*. Vol. 74, No. 2: 432–446.

Riker, William H. (1982). *Liberalism Against Populism—A Confrontation Between the Theory of Democracy and the Theory of Social Choice*. Long Grove, IL: Waveland Press.

Riker, William H. (1988). "The Place of Political Science in Public Choice," *Public Choice*. 57: 247–257.

Riker, William H. and Peter C. Ordeshook. (1968). "A Theory of the Calculus of Voting," *American Political Science Review*. Vol. 62, No. 1: 25–42.

Rose, Nikolas. (1999). *Powers of Freedom—Reframing Political Thought*. Cambridge, UK: Cambridge University Press.

Scott, James C. and Matthew A. Light. (2004). The Misuses of Numbers: Audits, Quantification, and the Obfuscation of Politics. In Jedediah Purdy, Anthony T. Kronman, and Cynthia Farrar. *Democratic Vistas: Reflections on the Life of American Democracy*. New Haven: Yale University Press, 115–137.

Shields, Christopher. (2016). Aristotle. *The Stanford Encyclopedia of Philosophy* (Winter 2016 Edition), Edward N. Zalta (ed.), plato.stanford.edu/archives/win2016/entries/aristotle.

Stedman Jones, Daniel Stedman. (2012). Masters of the Universe: Hayek, Friedman, and the Birth of Neoliberal Politics. Princeton, NJ: Princeton University Press.

Stehr, Nico. (2016). "Exceptional Circumstances: Does Climate Change Trump Democracy?" Issues in Science and Technology. Vol. 32, no. 2.

Stigler, George. (1984). "Economics: The Imperial Science?" *Scandinavian Journal of Economics*. Vol. 86, no. 3, 301–313.

Tullock, Gordon. (1972). Economic Imperialism. In James Buchanan and Robert Tollison (eds.), *Theory of Public Choice—Political Applications of Economics*. Ann Arbor, MI: University of Michigan Press, 317–329.

Van Horn, Rob and Philip Mirowski. (2009). The Rise of the Chicago School of Economics and the Birth of Neoliberalism. In Philip Mirowski and Dieter Plehwe (eds.), *The Road From Mont Pelerin: The Making of the Neoliberal Thought Collective*. Cambridge, MA: Harvard University Press.

Von Mises, Ludwig. (1949). *Human Action: A Treatise on Economics*. Auburn, AL: Mises Institute.

Williams, R.H. (1947). "Thyroid & Adrenal Interrelations with Special Reference to Hypotrichosis Axillaris in Thyrotoxicosis," *Journal of Clinical Endocrinology and Metabolism*. Vol. 7, 52–57.

4 Above Market and Democracy

The success of the business-funded campaign to turn the study of politics "scientific" was phenomenal. The understanding of politics as "the sphere of activity of a common that can only ever be contentious" (Rancière, 1999: 14) duly faded out from the discipline specializing in the study of politics. In its place came analysis of the market, allegedly apolitical and self-regulating, applied to the analysis of politics.

As neoliberal economists and political scientists were busy manufacturing theories and advertising the notion that the market treated *everyone* equally, scholars not lured by corporate money began to expose the reality that the neoliberal discourse tried to hide. As Galbraith noted, "A doctrine that celebrates individuality provides the cloak for organization" (Galbraith, 2007 [1967]: 270). He pointed out that there was a "great black hole of economics," deliberately constructed to prevent dominant narratives from noticing and discussing the prominent role of power relations in market economics and their social consequences (Galbraith, 1985: xxviii). In the power-free, self-regulating portrayal of the market, the firm and those within it were presented as "powerless automations wholly subordinate to market forces" (Galbraith, 1985: xxxv–xxxvi). As someone who had worked in the U.S. government and witnessed first-hand how the economy actually worked, Galbraith pointed out the discrepancy between the "reality" zealously sold by neoliberals and the actual reality: that big firms were far from powerless automatons. The economy no longer consisted of many small firms, all subordinate to the market. Instead, "a few hundred, perhaps a thousand or so, corporations" secured exceptionally dominant positions in their industries (Galbraith, 1985: xxxvii). The grave misapprehension, profusely supplied by dominant narratives in economics, concealed the fact that these firms took every feasible step to see that "what it decides to produce is wanted by the consumer at a remunerative price" and that "the labor, materials, and equipment that it needs will be available at a cost consistent with the price it will receive." In addition, powerful corporations must exercise control over "what is sold [and] what is supplied." In other words, they must "replace the market with planning." "As more time elapses and more capital is committed, corporations will be increasingly risky to rely on the untutored responses of the consumer" (Galbraith, 2007 [1967]: 27). Noting that it would be convenient to have a name for "the part of the economy that is characterized by the large corporations," Galbraith stated: "[o]ne is readily at hand; *Planning System*.[1] The planning system is the dominant feature of the New Industrial State" (Galbraith, 2007 [1967]: 12).

Whereas all businesses, big or small, would love to possess market-subordinating planning ability in order to minimize uncontrolled market influences and escape the need to participate in fair competition, only some firms *did* actually obtain such ability. To this fact, Galbraith pointed out, "industrial planning is in unabashed alliance with size," as the large organization "can tolerate market uncertainty as a smaller firm cannot" and "contract out of it as the smaller firm cannot." "Vertical integration, the control of prices and consumer demand and reciprocal absorption of market uncertainty by long-term contracts between firms all favor the large enterprise" (Galbraith, 2007 [1967]: 38). In fact, "[n]othing so characterizes the planning system as the scale of the modern corporate enterprise," which also explains why General Motors (GM), Exxon, IBM, and General Electric (GE) deviated so sharply from the legal image of a normal firm. The image of a firm as "an association of persons into an autonomous legal unit with a distinct legal personality that enables it to carry on business, own property and contract debts" (Guthmann and Dougall, 1948: 9, cited in Galbraith, 2007 [1967]: 90) was what a corporation should be, but not how corporations like GM, IBM, GE, and Exxon actually were. Unlike normal firms, these corporations were excessively influential in the markets and could not be said to be conducting business on equitable terms, as normal firms did, with other businesses. Galbraith went on to provide the evidence for his assertion: In 1976 the five largest industrial corporations had nearly 13 percent of all assets used in manufacturing, the 50 largest 42 percent, and the 500 largest 72 percent. The largest five employed 11 percent of the work force engaged in manufacturing, and the largest 15 employed 20 percent. Two corporations alone—American Telephone and Telegraph and GM—employed 2 percent of the total civilian labor force in the US. In terms of gross income, the three largest industrial corporations—Exxon, GM, and Ford—had a combined $125 billion in 1976. The gross revenues of Exxon, at $48.6 billion, were more than a hundred times the revenue of the US state of Nevada and more than three times that of the state of New York and about one-sixth that of the federal government of the United States.[2] "There is no evidence of any weakening of the trend either to larger and larger firms or to those having an ever greater share of the total output." As a result of planning, big corporations almost never lost money despite the "American business liturgy" that had "long intoned that this is a profit-and-loss economy." Between 1954 and 1976 there were only two years during which just five of the 100 largest industrial corporations lost money. All of the 100 recorded profits in seven of the 23 years. Between 1955 and 1976 all of the 50 largest merchandizing corporations made money in eight of the 22 years (Galbraith, 2007 [1967]: 92–4, 104).

In order to explain the way that large firms conducted planning, Galbraith coined the term "technostructure." Differing from the "management," the "technostructure" extended "from the most senior officials of the corporation to where it meets, at the outer perimeter, the white- and blue-collar workers whose function is to conform more or less mechanically to instruction or routine" and included all who brought specialized knowledge, talent, and experience to group decision-making. It was the technostructure, "not the narrow management group," that was "the guiding intelligence—the brain— of the enterprise" (Galbraith, 2007 [1967]: 87–8). Collectively, these highly trained individuals had a monopoly over crucial knowledge and took planning as their responsibility. Critically, the goal of planning was *not* to maximize profit, but to *control the market*. Only firms lacking the ability of planning and manipulation of market had no choice but to succumb to the market and find ways to maximize profit. For large firms, the "imperatives of technology and capital use do not allow the firm to be subordinate to the market." They must not only substantially control prices, but also "exercise influence on the amounts that are purchased and sold at these prices" (Galbraith, 2007 [1967]: 212). For these firms, in other words, the objective was not to explore market opportunities but to eliminate the uncertainties of the market—lack of wars for arms manufactures, lack of illness for pharmaceuticals, lack of inmates for privately run prisons, scientific discoveries linking products to health or environmental damages—by subordinating it to planning. It was something more important than its profits—namely, its *autonomy* from the market—that the technostructure was protecting.

Corporations having autonomy was a direct contradiction of the claim that the market was "self-regulating." The very point of claiming that the market was "self-regulating" was to establish that the market was the *only* entity possessing all the necessary information to allocate resources efficiently. This was a central reason for demanding that the state should keep its hands off market activities. It was also because the market was "self-regulating" rather than being controlled by any particular actors or organizations that competition in the free market could be deemed fair. Unsurprisingly, economists serving the interests of corporations insisted on profit maximization as the only goal of all businesses. "It is a far, far better thing to admit to monopoly profits, even at exploitive levels, than to concede that the market is impotent" (Galbraith, 2007 [1967]: 142). As a result, these economists asserted that profit maximization was "the strongest, the most universal, and the most persistent of the forces governing entrepreneurial behavior" (Stigler, 1952: 149, cited in Galbraith, 2007 [1967]: 142). They even made the claim that "few trends could so thoroughly undermine the very foundations of our free society as the acceptance by corporate officials of a social responsibility other than to make as much money for their stockholders as possible" (Friedman, 1962: 133, cited in Galbraith, 2007 [1967]: 142). The affirmation of such a corporate–society relationship with the corporation being the servant and the public being the master was necessary "for holding discussion of corporate behavior within the ambit of the economist" who was vigilant in dispelling any speculation that large corporations might be shaping social attitudes to their ends (Galbraith, 2007 [1967]: 159, 208). As a result of the diligent theorizing of the economist, society now firmly believed that the autonomy of the technostructure served a desirable social goal. No regulations threatening the autonomy of the industrial enterprise should be put in place, lest they interfered with the independent operation of the market mechanism to which large corporations were apparently subjected (Galbraith, 2007 [1967]: 211).

Viewed through the lens of this vital information, the whole picture of corporate–market–society–state relations that Galbraith presented looks very different from the half-picture that the neoliberals have insisted is all that there is to see. As Hayek, von Mises and Buchanan preached tirelessly about the sacrosanct self-regulating feature of the market, large corporations (including those who funded these economists, of course) succeeded in gaining increasing autonomy from the market through planning. Instead of remaining *within* the market to merely maximize profit like ordinary firms did, they sought to rise *above* it to allocate, semi-authoritatively, the resources of society for society. The self-regulating discourse disseminated by Hayek, Mises and Buchanan was absolutely essential for cloaking the doing of the technostructure (i.e., the manipulation of the consumer, the market, the state, and the way that the corporate–state–society relationship was perceived), in order to save the troublesome queries about accountability and resistance from society. As a result, autonomy from society and autonomy from the market went hand in hand, each depending on the other.

It was a dangerous oversimplification and grave misconception to assume even that a private sector existed, which was made up of corporations that were entirely separate from the public sector. Even those observant enough to point to the irregularity and illegitimacy of corporate influence on the state routinely failed to see that there was a significant overlap between the two, with the large corporation often being the one occupying the dominant position. "The mature corporation, so far from being separated organically from the state, exists... only in intimate association with it" (Galbraith, 2007 [1967]: 212). Unlike entrepreneurial corporations, which were not *dependent* on the state despite requiring public resources, favorable tariffs, and tax concessions, the mature corporation "depends on the state for trained manpower and the regulation of aggregate demand. These are important for the planning with which it replaces the market" (Galbraith, 2007 [1967]: 378–9). The unapologetic manner in which these corporations demanded and defended autonomy was thus remarkable, considering that much of their scientific and technical innovation came from, or was sponsored by, the state or by publicly supported universities and research institutions. Having acquired the fruit of publicly funded research for private use and with its autonomy secured, the technostructure then sent the state to take necessary measures to regulate aggregate demand for the products of the planning system and to bring society as a whole to underwrite the risk of its private investment: "In notable respects, the mature corporation is an arm of the state. And the state... an instrument of the planning system" (Galbraith, 2007 [1967]: 365). When planning takes place, the line separating government from the private corporation becomes very indistinct and even imaginary. "Each organization is important to the other; members are intermingled in daily work; each organization comes to accept the other's goals; each adapts the goals of the other to its own." The myth of separation, however, had to be maintained to protect the technostructure from "a good deal of awkward supervision" (Galbraith, 2007 [1967]: 385). For its part, the state also welcomed the diversion of potential accusations against partaking in economic planning that shifted risks of private investment to the public while retaining profits, market power, and political influence for big corporations. To further the "shedding" of "awkward responsibility," the myth of separation between government and the private corporation must be accompanied by the myth that the consumer is the real boss. In virtually all economic analysis, the initiative of economic activities was assumed to lie with the consumer. The flow was in one direction—"from the individual to the market to the producer" (Galbraith, 2007 [1967]: 263). This accepted sequence of the flow of economic activities protected the autonomy of the technostructure by outlawing a wide variety of public regulations and did so in the name of the individual. As a consequence, "government objection to lethal automobile design, disabling drugs, disfiguring beauty aids or high-calorie reducing compounds is [deemed as] interference with the individual's choice" (Galbraith, 2007 [1967]: 270). All this was affirmed by terminology that "implies that all power lies with the consumer. This is called consumer sovereignty" (Galbraith, 2007 [1967]: 263). What the doctrine that celebrated individuality was in fact providing, however, was "the cloak for organization." It was not the individuals' right to buy that was being protected, but "the seller's right to manage the individual" (Galbraith, 2007 [1967]: 270). It was the need to hide "the great producing organization which reaches forward to control the markets that it is presumed to serve and, beyond, to bend the customer to its needs" and "deeply influences his values and beliefs" (Galbraith, 2007 [1967]: 8) that propelled the indoctrination of individuality, consumer sovereignty, and a false sequence. Contrary to the characterization of private enterprise as subordinate to the market, the reality was a close fusion between the planning system and the state: "Members of the technostructure work closely with their public counterparts not only in the development and manufacture of products but in advising them of their needs" (Galbraith, 2007 [1967]: 478).

It would be a mistake to consider Galbraith as an opponent of economic planning. On the contrary, given the advancement of technology and the scale of capital needed to support technological development and production, he recognized not only the necessity of the planning system but also the autonomy of the technostructure as a functional necessity of the planning system. What he emphasized was the need to align the narratives with the reality, particularly given that the functions that the autonomy of the technostructure served could be positive ones beneficial to the broader society. When recognized as being part of the penumbra of the state, the mature corporation would be pressed harder to serve social goals and lose its ability to "plead its inherently private character or its subordination to the market as cover for the pursuit of different goals of particular interest to itself" (Galbraith, 2007 [1967]: 480).

For the same reason that Polanyi and Keynes failed to obtain corporate approval and a deserving place in mainstream economics teaching and research, Galbraith's warning against the neoliberal art of exclusion through inclusion was largely successfully contained. By the time that Galbraith published *The Culture of Contentment* in 1992, the neoliberal exclusion scheme was already so fruitful that up to 40% of the American population had become so marginalized that they no longer participated in the democratic process, "far less benefit from the growth in Society's wealth that the economic liberal revolutions of the 1980s brought in their wake" (Cockett, 1995: 333).

By the time that Galbraith published *The New Industrial State* in 1967, the mature corporations capable of sophisticated planning were already operating across national borders. The autonomy of the technostructure meant that the planning system was inherently antagonistic to the idea of national borders being the limit to the subordination of market to corporations. The technostructure did not, after all, put all of its efforts into ensuring the smooth running of supplier-serving order only to see foreign governments allowed to ignore or resist its meticulously designed planning. Hence, a corollary of planning was that multinational corporations became a force that encroached on "the role of the states as the sole regulators of external relations" (Kaiser, 1971: 712). The Marshall Plan was instrumental in transforming leading companies into large, transnational, and disproportionately influential corporations. For anyone accepting Polanyi's diagnosis for the breakout of the Second World War, it must be a terrible irony that the new seeds for dis-embedding the market were sown *while* war-torn Europe was being reconstructed. Among the multiple paths on which Europe could reconstruct itself, Europe picked, under overwhelming American influence, one that was "not quite that sought by the anti-fascist resistance, though fascist and Nazi collaborators were generally satisfied" (Chomsky, 1999: 9).

Conventional wisdom portrays generous help from the U.S. as indispensable for the postwar reconstruction of Europe. Some historians, however, questioned the macroeconomic and socio-political importance of the Marshall Plan for the recovery of Europe. Milward, for instance, argued that, using its own resources, Europe recovered very rapidly in 1945–46 and that "European reconstruction and unity were achieved almost despite the Marshall plan, not because of it" (Killick, 1997: 2–3). In the same vein,

Joyce and Gabriel Kolko depicted the U.S. government as "the servant of big business, aggressively exploitative, and expansionist." The goal of American foreign policy was therefore "to restructure the world so that American business could trade, operate, and profit without restrictions everywhere." The unanimity among American leaders in supporting this goal was absolute, and "it was around this core that they elaborated their policies and programs" (Kolko and Kolko, 1972: 2, cited by McKenzie, 2012: 111–2). Prior to the intervention of the U.S. to "help" Europe, self-contained national recovery in European countries featured various degrees of state- (as opposed to corporate-) dominated planned economy in the immediate postwar period. These programs, "in which local Communist parties participated, were judged (by the U.S. as) unsuited for maintaining capitalist rule in the long run" (Van Der Pijl, 2012: 148–9).

To rectify the mode of European recovery, "The administrators of the Marshall Plan approached Europe with clear ideas about the way in which economic and industrial life should be organized" (Killick, 1997: 156). Many in the administrative team complained that in Europe, "industrial and social structures and policies were still bound by inherited class attitudes and narrow national markets," and hoped that "both could be shattered by American leadership and experience." "They hoped to persuade Europeans to restructure their industry on the American pattern," with Volker Berghahn going so far as to claim that "World War II itself was about which model of capitalism would survive" (Killick, 1997: 157). Following the Marshall Plan blueprint, the U.S. "sharply terminated the period in which concessions had been made to state-monopolistic patterns of international trade and payments" through the implementation of the Plan. The state–monopolistic patterns were tolerated in the first place because it "allow[ed] the European states to stabilize class relations during the precarious end-of-war period, a tolerance arising from the absence of American capacity to underwrite an alternative policy." Once the Marshall Plan, which "[held] the keys to economic policy by its contribution to the modernization of the European economy," was drawn up, the state–monopolist structures were dismantled (Kees Van Der Pijl, 2012: 159).[3]

The composition of the Marshall Plan administrative team afforded evidence for Kolko and Kolko's depiction of the U.S. government as "the servant of big business." Averell Harriman, the U.S. Secretary of Commerce, was originally a railroad executive and Wall Street banker (Killick, 1997: 156). The Harriman Report was crucial in determining the Economic Cooperation Act of 1948. The goal of the Act was to encourage American businessmen to invest in Western Europe through investment guarantees, backed by government-negotiated treaties that protected U.S. investors from double taxation and prevented discrimination against U.S. capital.

It was the first such guarantee ever offered by the U.S. government to American business (Wilkins, 1974: 228).[4] Another key figure in the Marshall Plan administrative team, Paul Hoffmann, was not only the former president of a leading American car manufacturer, Studebaker, but also the founder of the "corporate-liberal vanguard," the Committee for Economic Development. Being within the power circle of European reconstruction, Hoffmann "intervened wherever policies contrary to the envisaged new production system threatened to be enacted, as for instance in the case of the original steel nationalization in Britain" (Kees Van Der Pijl, 2012: 149). In the course of 1947, with the Marshall Plan in place and "catered to the corporate-liberal fraction," U.S. corporate elites moved closer to power and resumed the tenure of "corporate-liberal internationalism" in Washington (Kees Van Der Pijl, 2012: 148).

With corporate-liberal internationalism at its core, the impact of the Marshall Plan on European society was sharply divided along class lines. On the one hand, it allowed "the liberal-internationalist bourgeoisie in Europe with a background in either the colonial or the Eastern European circuit of money capital to restructure their interests in a wider Pax Americana." As a result, in all Western European countries, "the liberal bourgeoisie strongly reasserted its position by entering the government or occupying key posts; but within other parties as well, notably Christian Democracy, the shift to liberalism was also manifest" (Kees Van Der Pijl, 2012: 161). On the other hand, European trade unions, "which were very strong in 1945, have generally been edged into a limited collective bargaining role similar to that of the unions in the USA." European national planning methods developed after the war were generally discredited under the influence of the American idea of meritocracy, with national institutions weakened while social services pared down to American levels (Killick, 1997: 156).

In the early 1960s the huge outflow of U.S. government money—or as Chomsky has put it, "generosity... largely bestowed by American taxpayers upon the corporate sector" (Chomsky, 1999: 9)—was replaced by a massive outflow of private direct investment to Europe as exemplified by the expansion of Ford at Dagenham in Britain. It was not attracted by the "dour statism of the late 1940s," but the recovery and integration of the European market (Killick, 1997: 175). The U.S. government itself acknowledged many years later the role that the Marshall Plan played in "'[setting] the stage for large amounts of private U.S. direct investment in Europe,'[5] establishing the basis for the modern Transnational Corporations" (Chomsky, 1999: 9). Although initially many American companies chose to invest in Britain, which had a common language, an increasing proportion went to Europe to take advantage of the Common Market after 1957. "These companies were often experienced at operating in several foreign jurisdictions and soon became, in some respects, more European than their various British, German, French and Italian competitors" (Killick, 1997: 176).

In 1969 in an article titled "The Giants' Causeway" *The Economist* stated that a main feature of the 1960s was "the spread of overseas investment in Europe and elsewhere by big American corporations." "More and more economic power," it pointed out, was falling into "the hands of a relatively small number of giant and truly international business corporations." The output of overseas subsidiaries of American corporations was estimated at about $120 billion a year in 1966 and $140 billion a year in 1969. The article pointed out, citing J.-J. Servan-Schreiber's French best-seller, *Le Défi Américain* (The American Challenge), that the numbers were "bigger than the gross national product of any free world country other than the United States; so that 'the third largest producing unit in the world, after the US and USSR' is formed by American subsidiaries abroad" (The Economist, 1969: 10). Moran further compared the growth rate of U.S. direct investment in Europe before and after the signing of the Treaty of Rome: Whereas the value of direct American investment grew from $2 billion to $4.15 billion at about the rate of 6% per year between 1946 and 1958, with the granting of European status to American subsidiaries located in the European Economic Community (EEC), the book value of direct U.S. investment jumped by 600% in 1967 compared with 1958 (Moran, 1976: 65–6). It is small wonder, then, that in addition to J.-J. Servan-Schreiber's *Le Défi Américain* there was a rash of books in the 1960s with titles such as *The Americanization of Europe* (Edward A. McCreary), *The American Invasion* (Francis Williams), and *The American Take-Over of Britain* (James McMillan and Bernard Harris), reflecting the dramatic presence of U.S. economic power in Europe (Wilkins, 1974: 345).

The U.S. reassured its partners by reducing, Buchanan style, the political question of "who controls" to a technical question of "what is the most efficient way of allocating resources?" American analysts declared "as a result of American investments, Europeans were enjoying better products, more competition, higher wages, and more rapid economic change than they would have otherwise" (Moran, 1976: 68). Instead of preventing further loss of control to corporations and reasserting the role of the state in protecting society, the European response to the "American challenge," the "Americanization of Europe," the "American invasion," and the "American take-over" was to join the U.S. in further dis-embedding the market from society. As a result, the European economy developed many "characteristic American features." Encouraged by European Community legislation, intra-European trade grew sharply, resulting in large-scale industrial restructuring that gave birth to firms that now operated throughout Europe. They were

> organized more like American oligopolies than traditional British or German firms; their managers use American methods, often learned in American-style management schools; their products and services are advertised in American-style media and are marketed in American-type stores. (Killick, 1997: 155)

Far from decreasing the dependence of European governments and societies from American investors, the growth of European multinationals aggravated the problem. At best, the efforts to avoid dependence upon the

U.S. "indicate a continuation of the drift that has characterized European integration since the late 1960s. At worst... they may lead to fragmentation and disintegration within the European Community" (Moran, 1976: 75–6). This observation, written four decades ago, rings poignantly true today.

In the late 1960s Charles Levinson, a key figure in international labor movement, stated that multinational corporations were "the first genuine world institutions with inherently global power and authority" (Levinson, 1969, cited in Cox, 1971: 582). In the structure of a global economy organized by a small number of corporations, "a separation between rich and poor [becomes] functional rather than geographical," rendering nationalism "the ideology of revolution on behalf of the weak and poor" and "transnationalism... the ideology of the dynamic rich."

> The rich of the world would be integrated transnationally wherever they are, while the poor remain marginal to the dominant system of production. The rich are those who seek their security in the corporation. The poor are those left outside its scope and for whom weak states are unable or unwilling to care—the marginal populations which cluster about the centers of industry, providing some supplies and services but not participating in the decisions or benefits of development. (Cox, 1971: 583)

The corrosive effect of the growth of multinational corporations on democracy was much more deliberate and pronounced than Cox's analysis suggested. Kaiser spelled out what Galbraith's book implied: Decisions made by these increasingly autonomous nongovernmental actors "affect the decision making context of governments, limit their maneuverability, and induce them to take measures vis-à-vis their own society or other governments" (Kaiser, 1971: 709). In the process, areas traditionally belonging to domestic politics were transferred to the realm of foreign policy, while restricted democratic control was justified by the alleged concerns over national security and interest. The intermeshing of decision-making in multinational frameworks typically resulted in a shift of political weight from legislatures to executives. The waiving of democratic scrutiny did not stop here. The executive could use the complexity of multinational decision-making rules to block undesired intrusions by parliament or public opinion. It could also insist that such negotiations had to be treated confidentially until they were concluded and that its liberty to disclose the content or state of progress was restrained, owing to the involvement of other governments. When supranational organizations such as the European Community were involved, it became even more difficult to locate responsibility when governments reached decisions by unanimity and majority vote (Kaiser, 1971: 713–5). The multinationalization of previously domestic activities such as investment, transfer of profits, or price policies could, "in the name of progress, efficiency, and interdependence," unavoidably undermine the Western systems of democracy. As a result, there was a pressing need to "preserve the primacy of politics without which there can be no democracy." Just as their national counterparts, transnational interests—and their links with national and international bureaucracies—must come under publicly exercised democratic control (Kaiser, 1971: 717–20).

None of Kaiser's warnings was heeded. Or, to put it more accurately, citizens of liberal democracies were never given the chance to reflect, let alone decide, upon what was imposed on them. The plain truth that the concentration of wealth was incompatible with democracy, a point on which the former U.S. Supreme Court Justice Louis D. Brandeis repeatedly elaborated in the early 19th century (Campbell, 2013), was not something that corporations would neglect to steer public attention clear of, in order to lower the guard of the public against what was actually taking place. Consequently, *as* wealth increasingly concentrated, citizens of liberal democracies became *increasingly* convinced that the free world operated exactly the way that multinational corporations skillfully sold it—well-functioning democracy coupled with an equivalently well-functioning and self-regulating market. As Galbraith made clear, the corporate-dominated planning system had no chance of persisting without the discourse that established in society a false sequence of economic activities, guided by consumer sovereignty. Also indispensable for the success of the planning system was the faith that the state was on the side of the people, guarding against the risk of wealth becoming concentrated in too few greedy hands. Protected by such discourse and beliefs, the act of subordinating the market, which depended on the simultaneous subordination of democracy, proceeded smoothly and quietly. By the 1970s the trails that the U.S. federal government had blazed for multinationals had expanded into a complex and comprehensive global structure best characterized as one where States and Corporations sat Above both Market and Democracy, henceforth the "SCAMD structure." Within the SCAMD structure, states and large corporations merged into one in managing and controlling societies, sharing the responsibility of maintaining the myth that market and democracy alike were functioning well and treating *everyone* equally.

Central to the subordination of market and democracy through the construction of the SCAMD structure was the role played by big oil companies, which were the "only truly aggressive U.S. investors abroad in the immediate post-war years," according to Wilkins (1974: 314), and the "archetype of the multinational corporation," according to *The Economist* (1969: 10). From 1946 to 1950 U.S. direct foreign investments grew by 143% in the petroleum sector, compared with 58% in manufacturing, 38% in mining, and 8% in utilities (Wilkins, 1974: 301). In 1945 General Douglas MacArthur established The Petroleum Advisory Group, which was "composed of leading U.S. and foreign oil men," implementing the mission to regulate and control the supply and distribution of petroleum "and ultimately to return the oil business to private ownership," an outcome that occurred in 1951. Under the aegis of the Group, the expansion of these oil companies entailed "every single facet of their business." With new Middle Eastern oil production—largely U.S.-owned—now opened up, shipping crude to the large European markets and refining it there proved immensely cost-effective for U.S. companies. Standard Oil of New Jersey, for instance, was not only able to operate 35 per cent *above* its prewar rate at its French refinery by 1948, but it also replaced its small refinery with a giant one at Fawley, Britain, rebuilt its refinery in Hamburg, Germany, and expanded its refinery in Italy (Wilkins, 1974: 315). Jersey Standard was not alone. Wilkins went on to document the expansion of Socony-Vacuum, Texaco, and Caltex in Europe in the late 1940s and early 1950s (Wilkins, 1974: 315). For the 1950–60 and 1960–70 periods the growth rates of U.S. direct investments in Europe in petroleum reached 410% and 310% respectively (Wilkins, 1974: 330).

The corrosive effect of the increasingly influential oil companies on democracy went beyond the drastic asymmetry of the power that such multinationals enjoyed in relation to other members of society. To begin with, it was not through random luck that big oil companies emerged as the archetype of democracy-eroding multinationals. According to Timothy Mitchell, the shift of energy dependence from coal to oil in the postwar period was closely linked to the role that coal had played in the rise of mass democracy. Having taken note of such a link, the US government, allied with large corporations, purposefully altered both America's and Europe's (through the Marshall Plan)[6] reliance on energy from coal to oil. To explain the relations between fossil fuels and democracy, Mitchell pointed out that:

> [t]he carbon itself must be transformed, beginning with the work done by those who bring it out of the ground. The transformations involve establishing connections and building alliances—connections and alliances that do not respect any divide between material and ideal, economic and political, natural and social, human and non-human or violence and representation. The connections make it possible to translate one set of resources and powers into another. Understanding the relations between fossil fuels and democracy requires tracing how these connections are built, the vulnerabilities and opportunities they create and the narrow points of passage where control is particularly effective. Political possibilities were opened up or narrowed down by different ways of organizing the flow and concentration of energy, and these possibilities were enhanced or limited by arrangements of people, finance, expertise and violence that were assembled in relationship to the distribution and control of energy. (Mitchell, 2009: 401)

When energy was more concentrated and less dispersed in terms of geography and handling, it gave workers the leverage to organize strikes, promulgate mass movements, and eventually meet the demands of modern representative democracy. In contrast, when the extraction, production, and distribution of energy relied more on machines, pipelines, and cross-oceanic oil tank ships, the menace that organized labor could pose was removed. The term "carbon democracy" was therefore used by Mitchell to refer to "the central place of coal in the rise of mass democracy, and then to the role of oil, with its different locations, properties and modes of control, in weakening the forms of democratic agency that a dependence on coal had enabled" (Mitchell, 2011: 143).

Much as big oil companies served as the archetype for multinational companies, the political and commercial constellation surrounding oil served as the archetype for the global SCAMD structure. In the process of shifting energy dependence from coal to oil, the West built many types of scaffolding for its imperial power in the Middle East through trusteeships, mandates, and development programs to manage the production and distribution of oil (Mitchell, 2009: 406). This trans- and supra-national planning system further consolidated the trans- and supra-national power structure, which then helped to create the perfect moment for the thorough expulsion of Keynes's economic theories from the policy world.

The 1973–74 oil crisis offered people in the West an unwelcome lesson on the laws of the market: When Middle East oil-producing countries cut the supply of oil in 1973, the price of oil soared as cars queued up at gas stations. This textbook case of the iron law of supply and demand brought Keynes' influence on economic policy-formation to an end. The canonical explanation for the dwindling of policy relevance of Keynes' theories was that the deplorable economic situation following the oil crisis had "proven" that not only did Keynesian solutions not work, but the policy of demand management through full employment had induced a period of "stagflation." The neoliberal effort to imprint this explanation on the minds of decision makers, intellectuals, and the public had been extremely successful. There were, however, a number of notable holes in the narratives that insisted that Keynes' theories failed to work and caused stagflation. For instance, the "Keynesian anti-inflation" solution prescribed at the time was, according to Davidson, Galbraith[7] and Skidelsky,[8] not based on Keynes's theories at all. There was a "charade of passing off an analysis that had nothing to do with Keynes as "Keynesian" (Davidson, 2009: 161). Keynes died in 1946, long before the neoliberal force grew substantial. When Milton Friedman forced a set of assumptions on Keynes' theories that Keynes would have objected to, and labeled policies derived from such a theory as "Keynesian," Keynes was not there to expose the neoliberal distortion of his theories.[9] Most relevant to *Surviving Democracy* however, among the holes in the story of Keynes's failed theories, was the role played by the sharp increase in oil price.

Mitchell began the chapter "The Crisis That Never Happened" by noting that "[t]he postwar petroleum order and the prosperity it brought seemed to collapse too easily" (2011: 173). He did not argue that the law of supply and demand in the so-called 1973–74 oil crisis was a fiction. He did believe, however, that it was "a piece of equipment carefully fabricated by certain parties to a dispute. To achieve their goals, those participants tried to organize an event that was assembled and performed in such a way that the laws of economics might operate" (2011: 174). The problem of using the crisis as an illustration of supply and demand lies in the fact that "it is difficult to know how much of the increase in the price of oil in the winter of 1973–74 was associated with a cut in supply, or even by how much the supply was cut" (Mitchell, 2011, 174). Not only was there evidence showing that the producing countries, taking advantage of higher prices, increased production after initial cutbacks and maintained the overall supply, but there was also great uncertainty about the price of oil. "For fifty years the oil companies had worked to prevent the creation of a 'market price' for crude oil." Consequently, "there was no place, publication or regular mechanism of exchange for determining the going price, so as the crisis unfolded no one knew what 'the market' was." Moreover, as "interruptions in the supply of oil from one source could be made up from another, the embargo against the United States 'never happened' " (2011, 175).

Echoing Galbraith's observation concerning the ability of mature corporations to subordinate the market, Mitchell noted that "[m]arket competition destroyed profits and ruined companies and had... to be prevented" (Mitchell, 2009: 408). The state–corporate trans- and supranational planning system, therefore, devised a long-distance machinery for maintaining oil supply scarcity that involved government quotas, price controls, consortium agreements on development of new oil discoveries, and "payments for *not* producing oil" (Mitchell, 2009: 408–9). Likewise, Rowley et al. saw the stagflationary process of the 1970s as resulting from the manipulation of oil prices by a complex structure of oligopolistic agencies rather than from economic policies informed by Keynes's theories. Moreover, they highlighted the profound effects that price manipulation and the resulting stagflation had on the distribution of national income and corporate profits. While companies that focused on domestic markets suffered, the degree of monopoly of major oil companies increased substantially. In addition, large banks emerged as winners, as they absorbed petrodollars and recirculated them to oil-producing countries and energy-related projects.[10] Meanwhile, major U.S. arms producers experienced a boom despite the sharp decline in military spending that followed withdrawal from Vietnam in 1975. These winners of the oil price spike—major armament, energy, and financial corporations—were the members of an "Armadollar-Petrodollar Coalition" and major beneficiaries of the Petrodollar recycling system. It was during this period that oil-exporting countries in the Middle East became the world's largest arms-importing countries (Rowley et al., 1989: 18–20). The crisis of 1973 was hardly the only "oil crisis" followed by booms in arms exports. Rowley et al. found that, throughout the late 1960s and 1970s, the interaction of arms imports and oil exports followed an almost stylized sequence. An increase of arms imports would spark armed conflict, which in turn facilitated the outbreak of an oil crisis. The "scarcity" created by the crisis would further increase oil revenues, which would then be used to fund further arms imports, initiating a new cycle (1989: 34). The positive relationship between "armaprofits" and "petroprofits" left Rowley et al. with the overwhelming impression that, "somewhere lurking behind the financial flows," significant individuals representing the large arms, oil, and financial companies must be jointly involved in shaping U.S. policies for the Middle Eastern region (1989: 42).

The U.S. government report on the U.S.-Saudi Arabian Joint Commission on Economic Cooperation served to confirm the suspicion of Rowley et al. The report stated that in the wake of the oil embargo and price spike, Saudi Arabia was left with "a substantial amount of petrodollars which could be used for development purposes." Accordingly, the Commission was established in June 1974 to assist "Saudi industrialization and development while recycling petrodollars, and facilitating the flow to Saudi Arabia of American goods, services, and technology" (U.S. General Accounting Office, 1979: 2). A more powerful piece of evidence confirming the impression of "significant individuals" "lurking behind the financial flows" came in the form of excerpts from the secret Bilderberg meeting[11] that took place five months prior to the announcement of the oil embargo. At the meeting held at Saltsjöbaden, Sweden in May 1973, an American participant presented a paper that foretold the "political and financial details of the oil price hike" (Zalloum, 2011: 123). The paper's presenter was Walter Levy, who headed the petroleum section of the Office of Strategic Services[12] in the Second World War and guided the petroleum sector of the Marshall Plan after the war (Chicago Tribune, 1997). In the paper Levy declared prophetically that the cost of Middle-Eastern oil "would rise tremendously, with difficult implications for the balance of payments of consuming countries. Serious problems would be caused by unprecedented foreign exchange accumulations of countries such as Saudi Arabia and Abu Dhabi." He also foresaw that "a complete change was underway in the political, strategic and power relationships between the oil-producing, importing, and home countries of international oil countries" (Bilderberg Meetings, 1973: 23). He then projected an imminent increase in the petroleum revenues of Middle Eastern oil producing countries from $9 billion in 1972 to $40 billion by 1980, (Bilderberg Meetings, 1973: 24) which would "translate into just over 400 per cent, the same level Kissinger was soon to demand of the Shah" (Engdahl, 2004: 131). The purpose of the secret Bilderberg meeting in that year was therefore to "manage the about-to-be-created flood of oil dollars, a process U.S. Secretary of State Kissinger later called "recycling the petrodollar flows" (Engdahl, 2004: 130). With the Bilderberg meeting, a group of powerful men[13] decided to launch "a colossal assault against industrial growth in the world, in order to tilt the balance of power back to the advantage of Anglo-American financial interests and the dollar" (Engdahl, 2004: 135).

As Mitchell pointed out, what was peculiar about the oil crisis was the fact that the postwar petroleum order had collapsed too easily. The framing of a sequence of events as a "crisis" simplified "changes in multiple fields, involving various agents, into a unique event, so that a single moment, with a single agent, appears responsible for a collapse of the old order." To understand the attempt to set back democratic politics, it was necessary to "follow changes in the multiple dimensions of the oil order, and the work that was done to simplify what happened into a crisis, for which an outside force—the Arab oil states—could be made culpable" (Mitchell, 2011: 173). Engdahl's analysis showed that the entire constellation of events surrounding the outbreak of the Yom Kippur War in October 1973 that triggered the embargo was "secretly orchestrated by Washington and London" (Engdahl, 2004: 136).[14] The fluctuation of prices in oil bore witness to the success of this massive transnational planning: Between 1949 and 1970 the average price for Middle East crude oil was approximately $1.90 per barrel. As the fateful Bilderberg meeting convened in early 1973 and discussed an imminent 400% future rise, the price rose to $3.01. "By January 1974, that 400 per cent increase was a fait accompli" (Engdahl, 2004: 138).[15]

Never in history had the destiny of the entire world been controlled by such a small circle of powerful men. The impact of the events known as the 1973–74 oil crisis went far beyond the tilting of the balance of power back to the advantage of Anglo-American financial interests and the dollar. The events fundamentally changed the way that economic theories were debated and economic policies formed, forcing Keynes' theories out of relevance. They demonstrated the formidable planning power of the SCAMD structure while in the process extracting further nutrition to consolidate and expand the structure. From that point on, the maneuvering room became ever smaller for market and democracy to work in ways that diverged from the wishes of the controlling elites.

Notes

1 Alluding to Polanyi's insight that "laissez-faire was planned, planning was not" (Polanyi, 1944: 147).
2 And about 2,700 times the revenue of a country like Sweden (Paxton, 1977: 1338).
3 In the same vein, Killick noted that "A major element of US policy in the Marshall Plan was to allow the Europeans to improve their trade balances by discriminating against American goods; this kept them in the (open) international market and on the US side in the Cold War. By the mid-1950s the European recovery convinced the Americans, with some pleasure, that their policy had been successful. Europe was not seen as a threat, and even the Treaty of Rome, which contained overtly protectionist measures and did raise some alarm, was generally accepted as valuable integration. It was not until the early 1960s that the USA began to complain strongly about EEC discrimination, but by this time the machinery was well-established" (Killick, 1997: 173).
4 Wilkins noted that the emergence of "U.S. and U.S.-supported international financial bodies" such as the Economic Cooperation Act, the European Investment

Bank, and the International Bank for Reconstruction and Development that "could assist U.S. private direct investment in foreign lands" was truly a "distinctive aspect of the postwar years" (Wilkins, 1974: 229).

5   U.S. Commerce Department, 1984 (Chomsky, 1999: 9).

6   As Mitchell noted, "the Marshall Plan paid for Europe to postpone plans to rebuild its battered coalfields and instead to purchase oil—supplied from the Middle East but paid for in US dollars" (Mitchell, 2009: 415).

7   "When *The New Industrial State* appeared, American Keynesianism was in its glory days.... And yet Galbraith saw that in their ascent, Keynes's doctrines had been distorted by his followers. The planning system had adapted demand management to its own purposes. Growth instead of full employment was now the paramount policy objective. Fluctuations in growth would receive immediate public policy attention; increases in unemployment would not unless they threatened the political stability of the system. The planning system contrived to favor growth just rapid enough to assure stable growth among its constituent firms, yet not so rapid that it would restore the power or perhaps the lost militancy of the unions. Growth generated by tax cuts might therefore lead toward full employment, but it might never get there" (Galbraith, James, 2007: xvi).

8   According to Skidelsky, postwar mainstream economists treated Keynes' economic theory as applicable only when money and wages were "sticky," meaning they were very slow to change or rigid. "Thus his theory was robbed of its theoretical bite" (Davidson, 2009: 163).

9   See the appendix titled "Why Keynes's Ideas Were Never Taught in American Universities" in Davidson 2009, especially 163–4. Milton Friedman was far from the only economist distorting Keynes's theories. Paul Samuelson, the most prominent Keynesian economist, also labeled ideas having nothing to do with Keynes as "Keynesian." "Samuelson's neoclassical theory became the foundation of what professors and students of economics believed was Keynes's theory. Accordingly, Keynes's revolutionary *General Theory* analysis was never adopted as part of mainstream economics. Consequently, in the 1970s, ... economists... such as... Milton Friedman..., easily defeated Samuelson's 'Keynesianism' on the grounds of the logical inconsistencies between Samuelson's neoclassical foundations and his 'Keynesian' economic policy prescriptions" (Davidson, 2009: 167–72).

10   This point was also emphasized by Engdahl: "[W]hile Kissinger's 1973 oil shock had a devastating impact on world industrial growth, it had an enormous benefit for certain established interests—the major New York and London banks, and the Seven Sisters oil multinationals of the United States and Britain. By 1974, Exxon had overtaken General Motors as the largest American corporation in gross revenues. Her sisters, including Mobil, Texaco, Chevron and Gulf, were not far behind." Meanwhile, banks such as "Chase Manhattan, Citibank, Manufacturers Hanover, Bank of America, Barclays, Lloyds, Midland Bank all enjoyed the windfall profits of the oil crisis" (Engdahl, 2004: 141). Galbraith also pointed out that, during "the most severe economic down-turn since the Great Depression," "[o]nly four of the largest one hundred industrial firms lost money in 1974; only three in 1975" (Galbraith, 2007 [1967]: 103–4).

11   See next chapter.

12   Forerunner of the Central Intelligence Agency.

13   Present at Saltsjöbaden in that Bilderberg meeting were the president of Royal Dutch Shell, Gerrit A. Wagner, the chairman of British Petroleum, Sir Eric Drake, and the director of British Petroleum, Sir Denis Greenhill. Other noticeable participants included E.G. Collado, vice-president of Exxon Corp., Robert O. Anderson of Atlantic Richfield Oil Co., George Ball of Lehman Brothers investment bank,

David Rockefeller of Chase Manhattan Bank, and a regular participant at the Bilderberg meetings, Henry Kissinger (Engdahl, 2004: 286–7).

14 "Kissinger effectively controlled the Israeli policy response through his intimate relation with Israel's Washington ambassador, Simcha Dinitz. In addition, Kissinger cultivated channels to the Egyptian and Syrian side. His method was simply to misrepresent to each party the critical elements of the other, ensuring the war and its subsequent Arab oil embargo.... The war and its aftermath.... were scripted in Washington along the precise lines of the Bilderberg deliberations in Saltsjöbaden the previous May.... Arab oil-producing nations were to be the scapegoats for the coming rage of the world, while the Anglo-American interests responsible stood quietly in the background" (Engdahl, 2004: 136).

15 The sudden increase in the oil price enabled companies such as British Petroleum and Royal Dutch Shell to produce oil in the risky North Sea at a profit.

References

Bilderberg Meetings. (1973). Available at: https://publicintelligence.net/bilderberg-conference-1973.

Campbell, Peter Scott. (2013). "Democracy v. Concentrated Wealth—In Search of a Louis D. Brandeis Quote," *Green Bag*. Vol. 16, no. 3.

Chicago Tribune. (1997). "Walter J. Levy, 86, Dean of U.S. Oil Economists," 15 December.

Chomsky, Noam. (1999). *The Umbrella of U.S. Power—The Universal Declaration of Human Rights and the Contradictions of U.S. Policy.* New York: Seven Stories Press.

Cockett, Richard. (1995). *Thinking the Unthinkable—Think-Tanks and the Economic Counter-Revolution, 1931–1983.* London: Harper Collins Publishers.

Cox, Robert W. (1971). Labor and Transnational Relations. *International Organization.* Vol. 25, No. 3, 554–584.

Davidson, Paul. (2009). *The Keynes Solution—The Path to Global Economic Prosperity.* New York: Palgrave Macmillan.

Engdahl, William. (2004). *Century of War: Anglo-American Oil Politics and the New World Order.* London: Pluto Press.

Friedman, Milton. (1962). *Capitalism and Freedom.* Chicago, IL: University of Chicago Press.

Galbraith, James K. (2007). Foreword to 2007 edition. *The New Industrial State.* Princeton, NJ: Princeton University Press.

Galbraith, John K. (1967 [2007]). *The New Industrial State.* Princeton, NJ: Princeton University Press.

Galbraith, John K. (1985). Introduction to the Fourth Edition: On the Perils and Rewards of Economic Dissonance. In Galbraith, John K. 2007 [1967]. *The New Industrial State.* Princeton, NJ: Princeton University Press.

Galbraith, John K. (1992). *The Culture of Contentment.* Boston, MA: Houghton Mifflin Company.

Guthmann, Harry G. and Herbert E. Dougall. (1948). *Corporate Financial Policy.* 2nd ed. New York: Prentice-Hall.

Kaiser, Karl. (1971). "Transnational Relations as a Threat to the Democratic Processes," *International Organization.* Vol. 25, No. 3, 706–720.

Killick, John. (1997). *The United States and European Reconstruction 1945–1960*. Edinburgh: Keele University Press.

Kolko, Joyce and Gabriel Kolko. (1972). *The Limits of Power: The World and United States Foreign Policy, 1945–1954*. New York: Harper and Row.

Levinson, Charles. (1969). Towards Industrial Democracy. Statement delivered to the First International Trade Union Conference on Industrial Democracy, Frankfurt-am-Main, Federal Republic of Germany, 28–29 November 1968. Reprinted in *ICF Bulletin*, January 1969, cited in Cox, 1971: 582.

McKenzie, Francine. (2012). Exporting the American Experience: Global Economic Governance and the Foreign Economic Policy of the Truman Administration. In *A Companion to Harry S. Truman*.

Mitchell, Timothy. (2009). "Carbon Democracy," *Economy and Society*. 38(3): 399–432.

Mitchell, Timothy. (2011). *Carbon Democracy—Political Power in the Age of Oil*, London: Verso Books.

Moran, Theodore H. (1976). "Multinational Corporations and the Political Economy of U.S.-European Relations," *Journal of International Affairs*. Vol. 30, No. 1.

Paxton, John (ed.). (1977). *Statesman's Yearbook 1975–1976*. London: Macmillan.

Polanyi, Karl. (2001 [1944]). *The Great Transformation: the Political and Economic Origins of Our Time*, Boston, MA: Beacon Press.

Rancière, Jacques. (1999). *Disagreement—Politics and Philosophy*. Minneapolis, MN: University of Minnesota Press.

Rowley, Robin, Shimshon Bichler, and Jonathan Nitzan. (1989). "The Armadollar-Petrodollar Coalition and the Middle East. Department of Economics Working Paper," 10/89. McGill University.

Stigler, George J. (1952). *The Theory of Price*. New York: Macmillan.

The Economist. (1969). "The Giants' Causeway," 27 December: 10.

United States General Accounting Office. (1979). "Report by the Comptroller General of the United States. The U.S.-Saudi Arabian Joint Commission On Economic Cooperation," ID-79-77. 22 March. www.gao.gov/products/ID-79-7.

Van Der Pijl, Kees. (2012). *The Making of An Atlantic Ruling Class*. London: Verso Books.

Wilkins, Mira. (1974). *The Maturing of Multinational Enterprise—American Business Abroad from 1914 to 1970*. Cambridge, MA: Harvard University Press.

Zalloum, Abdulhay Y. (2011). *America in Islamistan: Trade, Oil and Blood*. Bloomington, IN: Trafford Publishing.

5   EU, The Poster Child?

The EU is often praised as the poster child of climate change mitigation (EURACTIV, 2017; Groenleer and van Schaik, 2007; Gupta and Grubb, 2000; Oberthür and Pallemaerts, 2010; Oberthür and Roch Kelly, 2008; Schreurs and Tiberghien, 2007). Although it is among the largest historic emitters,[1] according to popular narratives, the EU has taken the lead in addressing the crisis, particularly compared with the U.S. This framing of the EU as a leader in the climate crisis, however, ignores the active role that the corporate-serving "ever closer Europe" has played in causing the crisis. Instead, the impression is given that climate change "just happened," and that the EU reacted to an external, exogenous threat. For both climate and socioeconomic distributive rules, systems are highly sensitive to initial conditions. In Europe, the EU sits at the crucial intersection of these two systems, and the effects of European integration reverberate far beyond Europe. Before describing the failure of the EU's carbon-trading scheme—the backbone of its climate action—in emission reductions, and the success of the scheme in rewarding big polluters with big money, this chapter first advances the argument that the EU's *raison d'être* is incompatible with honest, timely, and effective climate mitigation. An examination of the key actors behind the single market and single currency agendas reveals that the core mission of the EU is to entrench the neoliberal socioeconomic order. The fact that big polluters played a dominant role in constructing the essential structures and institutions of the EU makes mitigation even less compatible with the EU's unannounced goal of protecting upward redistribution. Unless the EU resets its core mission and embraces a vision fundamentally different from those of its creators, its climate action will continue to consist merely of endless tinkering, precipitating the process of climate change.

The genesis of the EU can be found in two closely linked elite groups—the Bilderberg group and the European Round Table (ERT) of Industrialists—with overlapping members. In a significant way, it was members of these groups that authored the Single European Act (SEA) and the Treaty of Maastricht. These treaties serve the same function as did the constitution that neoliberals helped to write for Chile, namely, locking in rules that ensured a continued widening of wealth- and power-asymmetry between large corporations and ordinary citizens. The Bilderberg group, which brings together global—but mainly transatlantic—business and political elites in secrecy, is a manifestation of the way that states and corporations operate above both market and democracy, just as Galbraith, Kaiser, Mitchell, Rowley, and Engdahl described (see Chapter 4). The interests of petrochemical giants have been well represented in the group from the start. As Skelton put it, "[t]he veins of Bilderberg run with oil, [with] its beating heart [being] the Dutch royal family and its oil interests" (Skelton, 2018).[2] This oil-rich elite group, described as "the birthplace of the European Community" (Gill, 1990: 132, citing Hatch, 1962: 240),[3] was instrumental in subordinating European democratic systems under neoliberal rules and constitutionalizing such an arrangement. The Bilderberg Group was formed in 1954 at the direct behest of Joseph Retinger, also the founder of the European League for Economic Cooperation (ELEC) (Nollert & Fielder, 2000: 189). The ELEC was a free-trade promoting association of conservative European elites that shared "the laissez-faire spirit of the International Chamber of Commerce with which it had many personal links" (Van Der Pijl, 2012: 120).[4] The Bilderberg initiative coincided with the creation, in September 1953, of the Coleman Committee in the U.S., tasked with promoting trade liberalization. Its members included prominent industrialists, lawyers, and bankers dedicated to free trade, and many soon became deeply involved in the Bilderberg group as well (Aubourg, 2003: 95). The inaugural Bilderberg meeting was held in Hotel de Bilderberg in Oosterbeek, Netherlands in May 1954 (Nollert & Fielder, 2000: 190; Engdahl, 2004: 135). Much of the funding for early Bilderberg Group meetings came from the Central Intelligence Agency (CIA), Unilever, and the Ford Foundation (Aldrich, 1997: 216; Aubourg, 2003; Wilford, 2003: 76–7; Van Der Pijl, 2012: 183).

Conspiracy theories suggest that the Bilderberg Group is a power center forging a "New World Order" aligned with the interests of the world's richest and most influential elites. Retinger dismissed such speculation, stating that "[a]t our meetings…., we draw conclusions, but there is no voting on resolutions" (1956). Viscount Etienne Davignon,[5] a former chairman of the group and a central figure in the establishment of the EU, similarly insisted that, while the group is the place where people with influence meet other people with influence, things happen in a "much more incoherent fashion." Participants do agree that "it is wrong not to try to deal with a problem," but there is no concrete "action plan containing points 1, 2 and 3…"(BBC, 2005).[6]

Whether the Bilderberg meetings constitute a conspiracy is quite beside the point. The problem lies in the discrepancy between the nurtured belief that market and democracy treat *everyone* equally, and the reality. If the market is truly self-regulating and the democratic political system is truly bottom-up, as the SCAMD (States and Corporations sitting Above both Market and Democracy) elites insist, what purpose does the stringently secretive socialization of the wealthiest and most powerful serve? As *The Times* reporter Moorehead pointed out in 1977, it is, after all, the *members* of the Bilderberg meetings that makes them so important. They are government ministers, chairmen of big banks, and leading industrialists. The fact that these busy and powerful people commit themselves, year after year, to devote three days to the conference is itself proof of its importance. It is no accident that 95% of the attendees belong to what is loosely termed "the establishment." As one participant rhetorically asked, "How else do you attract the powerful, other than with other powerful people?" (Moorehead, 1977)

Climate change would never have spun out of control, had powerful individuals not escaped democratic accountability and systematically made far-reaching collective decisions at places like Bilderberg. Retinger boasted the merit of secrecy for conversations held at Bilderberg. Unlike official international meetings attended by "representatives accompanied by a retinue of experts and civil servants," Bilderberg enabled participants to "hold frank and full discussions..., smoothing over difficulties.... and... finding a common approach." More importantly, "[e]verybody who attends our meetings does so in his private capacity even if he is the leader of a government..., and thus he is not responsible to his supporters for anything he may say" (Retinger, 1956). In 1977 Moorehead concluded that the Bilderberg group "may try to pass themselves off as a steering group for democracy, the senior common room of the industrial West. They are, in fact, heavily biased towards politics of moderate conservatism and big business" (Moorehead, 1977).

Having such an oil-rich secretive group as its birthplace, the European Community (EC) was inherently anti-climate, as well as anti-democratic.[7] The advancement of anti-democratic structures and processes within the foundations of the EC (and later the EU) was indivisible with the advancement of anti-climate structures. As early as 1955, participants began to express their concerns over "the dangers inherent in the present divided markets of Europe." It was generally recognized that "it is our common responsibility to arrive in the shortest possible time at the highest degree of integration, beginning with a common European market." The kind of common market envisioned by a group like this was predictable. A U.S. participant asserted that the success of economic performance in the West was "validating the claim that economic systems, driven increasingly by the accumulated decisions of individuals and businesses, could achieve self-regenerative economic activity at high and acceptable levels to populations." He also stated that "we have had to consider what were the realistic levels of employment and unemployment." A European speaker expressed the need to create "a common currency" and "a central political authority" in Europe.[8] By 1958, the year in which the European Economic Communities (EEC) came into existence, "European monetary policy" was already a topic that the Group had placed on its meeting agenda (Bird, 1992: 472).[9]

The reason that the Bilderbergers persistently pushed for a single market and a single currency was the effect that such arrangements could create in widening corporate-citizen wealth- and power-asymmetry and closing off options for society. By bypassing national democratic scrutiny and reducing society's say in what corporations could or could not do, supranational market integration helped to expand the reach of corporate grasping. Eventually, this EU-nurtured asymmetry played a critical role in the struggle against climate change. By redistributing resources to large corporations, elite-dominated European integration empowered big polluters to engage in even more public relations (PR) campaigns to suppress awareness and delay resistance. Even after citizens saw through corporate PR tactics and began to demand change, crucial decision-making power had long been handed over to Brussels, where corporate lobbyists rather than citizens call the shots. "Regulatory capture," which has so plagued democratic countries across the world since the late 20th century, is hardly the right expression for what is taking place in the EU, where there is nothing to "be captured." Unlike nation states that have *demos* and are founded for reasons other than serving corporate interests, the EU owes its existence to far-sighted corporate elites and therefore does what its creators programmed it to do automatically and naturally without having to "be captured" first.[10] Not only was the blueprint of European integration authored by big polluters, together with other powerful industries, but corporate elites also formed "watchdog group" to ensure that transparency and democratic accountability did not interfere in the process and tamper with their meticulous design.

Before delving into the conversion of elite wishes into concrete institutional design in Europe, a little detour to Chile, the neoliberalizing experiences of which served as a template for the upcoming engineering of the EU, can be illuminating. The Chilean case is a manifestation of the construction of a sophisticated mechanism that tied the hands of the people whom the elites anticipated would, in the future, rebel against the exclusionary grasping. It was an extraordinarily effective mechanism that would later be copied and is now at work to prevent people in Europe and beyond to resist the mitigation-sabotaging force. In 1970 Salvador Allende, a Marxist activist, was elected by the Chilean people as the president of the country. Kissinger, then U.S. National Security Adviser and a key member of the Bilderberg group, said "I don't see why we need to stand idly by and watch a country go Communist due to the irresponsibility of its own people." In the following years, the CIA and large U.S. corporations destabilized the country with the aim of "making the economy scream" (Burbach, 2006). On 11 September 1973 the Chilean military ousted Allende in a U.S.-backed coup. While in the years that followed, both Milton Friedman through his advice on monetary policy, and Hayek through his anti-democratic advice to the junta constituted harms to the country, these harms cannot be said to be permanent. "The same," says MacLean, "cannot be said of James Buchanan, [whose] impact is still being felt today." Buchanan, having always emphasized the importance of the "rules of rule-making" and constitutionalizing the neoliberal order, guided the head of Chile's military junta, Gen. Augusto Pinochet to "arrange things so that even when the country finally returned to representative institutions, its capitalist class would be all but permanently entrenched in power." "Whereas the U.S. Constitution famously enshrined 'checks and balances' to prevent majorities from abusing their power over minorities," the Constitution Buchanan helped to design for Chile "bound democracy with 'locks and bolts'" (MacLean, 2017: 155). Neoliberals' early victory in Chile marked the beginning of a familiar pattern. Future inter- and supranational institution-building endeavors followed the same logic of protecting and entrenching the neoliberal order behind an impassable moat (2017: 157), such as was the case in the establishment of the EU.

In providing his "expertise" to the junta, Buchanan explained that the constitution needed "severe restrictions on the power of government." Keynesian deficits that spent government money on those who did not matter should no longer be tolerated. To fulfill this goal, the country must have a constitution that required a balanced budget. In addition, the independence of the Central Bank must be enshrined in the constitution. Finally, a requirement of high-threshold supermajorities must be in place to prevent any change of substance, making the constitution "a virtually unamendable charter." In the words of the leading U.S. historian of the Pinochet era, Steve Stern, it "promised a democracy protected from too much democracy" (MacLean, 2017: 158–61).

The anti-Keynesian balanced-budget fiscal policy and the blindly anti-inflationary monetary policies, secured through "independent" central banks, form the core foundation of the neoliberal economic program. Together, they constitute by far the most influential masterpiece of the neoliberal art of exclusion through inclusion. They serve as powerful instruments for protecting the savings of the rich, while depriving governments of the authority to make policies responsive to social pains, including (but of course not limited to) climate change. At least as early as 1981, the desire of European economic and political elites to apply the template of the Chilean constitutional design on Europe was already apparent. At the European Council meeting of March 1981, the heads of state agreed unanimously that, in order to reduce the level of inflation and stimulate growth, "prudent monetary policies, a healthy budgetary management, and the reorientation of public and private expenditure in the direction of productive investment" were essential (The European Council, 1981). By 1982 the neoliberal EU state-building had started, with the Commissioner for Industrial Affairs and Energy, Viscount Davignon, pushing hard for the creation of "a genuine single market for industry in the Community—e.g. European standards to replace national standards…, the 'European company' taxation which is neutral in its impact on competition, [and] measures to simplify customs formalities" (Davignon, 1982: 15). Viscount Davignon, the chairman of the Bilderberg Conference between 1999 and

2011, was involved in the group long before he became the Industrial Commissioner of the EC in 1977.[11] As the Industrial Commissioner, he worked tirelessly in converting the wishes of the Bilderberg elites into reality through the restructuring of the EC. Together with Pehr Gyllenhammar of Volvo,[12] Davignon created the ERT, a permanent elite group representing the bulk of large-scale European multinationals with which the Commission engaged in formal dialogue starting in 1983 (van Apeldoorn, 2002: 85).

Davignon and Gyllenhammar modeled the group on the U.S. Business Roundtable, which consisted of CEOs from the 200 largest US-based companies and had direct access to members of Congress and the executive branch in promoting their legislative agenda. In the Commission's Berlaymont building in Brussels, the first list of members to be invited to the ERT was drawn up in 1982 (Cowles, 1995: 504). The group was not meant to be just a high-powered lobby organization. Rather, the goal was to fundamentally change the way that the market operated in Europe through the promotion of the internal market. In the new single market, businesses could break free from domestic constraints and be provided with alternative paths of job and wealth creation, giving them more flexibility in labor decisions—including laying off workers and closing down factories—to restructure their organizations across Europe. On 6–7 April 1983, the ERT held its first meeting, with the President of the European Commission, Francois Xavier Ortoli, and Commissioner Davignon participating in all but the final session (Cowles, 1995: 505). Of the 17 initial participants of the ERT,[13] Philips, Fiat,[14] and Unilever were all founding members and later frequent participants of Bilderberg.[15] Shell was, because of Prince Bernhard, in the same rank. Jacques Solvay of Belgium's Solvay Chemical, who campaigned energetically, together with Wisse Dekker, the CEO of of Philips, for the unified internal market, was also a frequent Bilderberger (*The Economist*, 1988: 6). This public-private joint endeavor (involving the Bilderberg Group, the ERT, and the Commission) of the engineering of a top-down system, which locked in an upward re-distributional mechanism, is a quintessential example of the way the SCAMD structure works.

The ERT finalized a memorandum in 1983, emphasizing the importance of the European market serving as the unified home base for European firms to develop as powerful competitors in world markets. Pledging their determination to promote new wealth creation in Europe, the industrialists emphasized that "we cannot do this unaided... we need supporting political action." In view of the necessary "painful restructuring operations," the ERT members highlighted the imperative of installing a *"political-legal framework* within which these initiatives could be taken" (Cowles, 1995: 506–7). On 11 January 1985 Dekker,[16] a prominent member of the ERT, put forward a plan for the creation of an internal market entitled "Europe 1990—An Agenda for Action" (Dekker, 1985). The Dekker Paper set a clear timetable for steps to be taken before the goal of a single European market could be achieved, which the Paper urged should take place no later than 1990. On the very next day, the new Commission President, Jacques Delors, published the Program of the Commission for 1985 (Commission of the European Communities, 1985a).[17] It sang in such extraordinary unison with the Dekker Paper that little room was left for different voices to articulate alternative ways forward. At the European Council meeting in late March, the heads of state and government endorsed the Program and called for the creation of a single market by 1992, a move with unfortunate democratic, distributional, and ecological consequences.

With the endorsement of member state governments secured, Delors invited the ERT to the Commission to discuss ERT goals in June. ERT members met the Delors Cabinet and other "backroom boys" (key Commission officials) on several occasions to prepare for the meeting. In addition, Gyllenhammar met Delors privately to discuss the "official" ERT-Commission meeting on 14 June, which was also the date on which the landmark document—the Commission White Paper, "Completing the Internal Market," authored by Commissioner Lord Cockfield—was finalized (Cowles, 1995: 515; Commission of the European Communities, 1985b). At the Milan summit at the end of June, the heads of state and government endorsed the White Paper. Contained in the White Paper was the famous concept of mutual recognition and the nearly 300 pieces of legislation needed for the creation of a single market (Cowles, 1995: 516). Based on this blueprint drawn up collectively by the ERT and the Commission—both closely linked to the oil-rich Bilderberg Group—the EC member states signed the SEA in February 1986 and committed to creating a single market by 1992. Honoring the principle of mutual recognition which the Commission White Paper had promised the business community, the SEA bound member states to the principle that standards prevailing in one country must be recognized by the others as sufficient. The principle of mutual recognition sounded deceivingly innocuous. Leaving the impression that it did not intend to intrusively "substitute Community versions for existing national regulatory systems," it nonetheless outlawed "any impact of the latter on the free movement of commodities, services and factors of production" (Grahl and Teague, 1989: 34). The brilliance in this market completion tactic was that it bypassed the need to replace incompatible national regulations with harmonized European standards. A Commission report[18] even applauded the "superiority of this new, more ready way of integrating markets which will stimulate a healthy rivalry between the various national systems" (Grahl and Teague, 1989: 40). In the end, the design automatically weakened the legal and administrative controls which were asserted, even in the absence of evidence, to be crippling the corporate sector (Grahl and Teague, 1989: 40), creating a "race-to-the-bottom" dynamic (Schmidt, 2007: 667).

The famous decision by the European Court of Justice (ECJ) in the *Cassis de Dijon* (1979) verdict was repeatedly cited as a precedent for justifying the principle of mutual recognition.[19] As Alter and Meunier-Aitsahalia explained, however, the ECJ's *Cassis* decision did *not* contain the mutual recognition meaning that the Commission insisted it did. Instead, through the verdict, the Court "created a general 'rule of reason' whereby any national law with reasonable policy goals... would be tolerated" (Alter and Meunier-Aitsahalia, 1994: 539–40). Hence, contrary to the assertion of it being a precedent for mutual recognition, the judgment should be seen as a "triumph for regional and national specialties" and a setback for those who would "standardize purely for standardization's sake," for it gave full freedom to products lawfully manufactured and sold in their country of origin, and as a result better protected the consumer by giving them a wider variety of choice (Gormley, 1981: 459). The Commission activism in alter-ing the meaning of the ruling tremendously facilitated the SCAMD elites' pursuit of the single market agenda. In fact, before the ECJ ruled on the *Cassis* case, "the Commission had, for some time, been preparing to increase its attacks on barriers to trade within the Community" by 1978 (Gormley, 1981: 455). Once the verdict came out in 1979, Commissioner Davignon, a Bilderberg insider, pressed for an overhaul of European inte-gration based on a twisted interpretation of the ruling. By July 1980—before the creation of the ERT and the emergence of the Dekker Paper—the Commission already included the principle of mutual recognition of goods in its program, stating that "any product lawfully produced and marketed in one Member State must, in principle, be admitted to the market of any other Member State... [which] may not take an exclusively national view-point" (Official Journal No. C256, October 3, 1980, cited in Alter and Meunier-Aitsahalia, 1994: 541). The role of the Commission, or at least the part of the Commission under Davignon's control, as an agent of the oil-rich Bilderberg Group was revealing.

By any non-neoliberal standard, for polluter-filled groups such as the ERT to form a watchdog group to "monitor developments, identify... delays and apply pressure to government leaders" to keep momentum behind the implementation of the SEA (Cowles, 1995: 518) must be pre-posterous. By December 1986 the watchdog group—International Market Support Committee (IMSC) consisting of prominent Round Table mem-bers—was established. It held private meetings with the EC and national top-level officials, including the head of state or government. In such meetings that were "supported by thousands of contacts on *ad hoc* basis," (Van Apeldoorn, 2019: 142) government officials took note of "the ERT's suggestions and concerns regarding developments and overall strategies of the European Community" (Cowles, 1995: 519). The tone in which the IMSC communicated with leaders of member states was exemplified by its press release issued at the 24 June, 1987 EC Summit wherein it menaced: "show political will, or European industry will invest elsewhere" (Cowles, 1995: 519). Obediently, European leaders showed their political will. By 1988 the completion of the single market program became irreversible under the German presidency of the EC (Cowles, 1995: 520).

The rhetoric of low competitiveness that European companies had suffered relative to their American and Japanese counterparts[20] was what buoyed the SEA. It is important to remember, however, that the environment in which this "Eurosclerosis"[21] rhetoric spread was one wherein the nationality of a company was decreasingly relevant. By the end of the 1960s *The Economist* already saw the trend that "the [transnational] giants will grow smaller in number but much larger in size... [with] mergers... spreading rapidly over several frontiers." As a result, the question of whether "the new giants should be called British or German or Dutch or French or American might literally come to have no meaning" (*The Economist*, 1969: 10). In sum, the expressed goal of producing an economic union aimed at increasing the competitiveness of European firms relative to their North American and East Asian counterparts conveniently hid the fact that "much of what is normally referred to as 'European' capital involves US, Japanese and other transnational firms" (Gill, 1998: 10). Rather than competitiveness in European industries, therefore, it was the power- and wealth- asymmetry between corporate elites and the rest of society that the SCAMD elites pursued. This background information solves the puzzle that, despite the claim that an internal market was an imperative for European companies to outcompete the U.S. and Japanese companies, the establishment of the internal market actually *enhanced* external trade liberalization, with individual EU member states unilaterally abolishing over 6,300 quantitative restrictions against imports from third countries between 1990 and 1995 (Hanson, 1998: 59).[22]

The corporate takeover of roles previously performed by governments would not have been possible, had the SCAMD elites not honed their PR skills for achieving exclusion through inclusion via venues such as Bilderberg. The 1971 Bilderberg meeting offers a glimpse of how much the later materialization, through the single market project, matched early visions of Bilderbergers. At that meeting, Giovanni Agnelli of Fiat, later a prominent member of the ERT, noted that business has "*bypassed political reality,*" resulting in its having a great "social responsibility." Business must "[receive] from society... the *mandate* to allocate the financial, human and natural resources of society" (1971: 21, 23; emphasis added). As a result, he saw the need to "make considerable *investments* in the *management of society* and culture" (1971: 18; emphasis added). He emphasized that business must cooperate with the goal of "[devising] the ways whereby business can cooperate openly, without this cooperation being considered the unwarranted interference of a pressure group" (1971: 18). This self-appointed role for corporations to collectively govern society, without being held democratically accountable, was precisely what the neoliberalized structures of the EU were set up for.

The completion of the single market, however, was only a first step for fulfilling the goals set by the SCAMD elites. The crucial component in the neoliberal economic program—that of a single currency with a unified monetary policy monopolized in the hands of an independent central bank, as the Chilean Constitution had accomplished—was still missing.

Once again, it was the Bilderberg group that laid the major ground work necessary for eliminating socially responsive monetary policies of individual countries that could negatively affect the asset value of the wealthiest and filled the void with the European Central Bank. Once again, it was fundamentally anti-climate interests that drove the agenda and laid those foundations. At age 77 and as vice-chairman of Belgian energy firm Suez-Tractebel, Davignon commented on the financial crisis in March 2009, stating that the upcoming Bilderberg conference in that year, which he chaired, could "improve understanding on future action, *in the same way it helped create the euro in the 1990s.*" He added that "When we were having debates on the euro, people [at Bilderberg events] could explain why it was worth taking risks and [those who disagreed] had to stand up and come up with real arguments" (Rettman, 2009; emphasis added).

By the late 1970s macroeconomic policies in European countries had already gradually converged towards restrictive monetary policies and budgetary consolidation as a result of the diligent neoliberal campaign to push for the abandonment of Keynesian economic policies. When the European Monetary System (EMS) was created in 1979 to stabilize exchange rates and facilitate trade among EC currencies, fighting inflation to preserve the value of savings and assets was already an implicit goal. As a result, the EMS, which required formal withdrawal from commitments if a country decided to pursue expansionary policies, worked as a constraint on domestic politics that pushed toward more restrictive macroeconomic politics than would otherwise have been the case (Sandholtz and Zysman, 1989: 112). The EMS was nonetheless a long way from a single currency that could justify the creation of an independent central bank that wielded monetary disciplining power in the way that the Chilean central bank could. As the completion of the single market mandated by the SEA approached, Delors, emphasizing the incompatibility of capital liberalization and the EMS, announced in 1990 that: "The economic advantages of the Single Market are certainly not fully achievable without a single currency" (Talani, 2003: 125; 142). In 1992 the Treaty of Maastricht, which contained provisions needed for establishing a single currency in the European Monetary Union (EMU), was signed. Sharing with the Chilean constitution the goal of binding governments with locks and bolts against ever being able to pursue Keynesian expansive policies, the treaty limited government deficits to 3% of GDP and public debt levels to 60%. In essence, the Maastricht Treaty:

> creates a body of Platonic monetary guardians, accountable to no one, to frame and execute one of the most important aspects of policy in modern economies, affecting hundreds of millions of people. This was done in the name of insulating monetary policy—and its primary objective of price stability—from political pressure, and of endowing the new European central bank with political independence. (Cooper, 1994: 70)

In 1997 the Stability and Growth Pact was born to monitor the enforcement of the Maastricht deficit and debt limits, ensuring that member states stayed on the track of sound public finances. Hence, reminiscent of the Chilean constitution, the Treaty served as a way of "locking in" political commitments to orthodox market-monetarist fiscal and monetary measures.[23] Through a rules-based economic constitution that mandated tight monetary and financial discipline, inflation would be kept low to protect savings. As an instrument for safeguarding property rights and investor freedoms, this constitutional framework separated "the economic and political in ways that lessen the possibility for democratic accountability." It bound governments to be "more responsive to the discipline of market forces and correspondingly less responsive to popular-democratic forces and processes" (Gill, 1998: 5; 9). The procedure of locking democracy in chains was so skillfully executed through sophisticated PR work that it encountered little popular resistance. These chains that have prevented democracy from performing its duty are the very reason that timely, honest, and effective climate mitigation measures find no place in today's neoliberalized world.

Similar to the way governments of advanced democracies handled climate-related knowledge, in their pursuit of the European single currency, member state governments ignored the warning coming from economists that the monetary union lacked theoretical and empirical grounding. On 12 June 1997 an open letter signed by 331 economists was sent to the heads of government and state, which stated:

> Your economic advisers have told you that the EMU, as laid out in the Maastricht Treaty and further regulated in the Dublin Stability Pact, will bring Europe more jobs and prosperity. We... are afraid that the opposite is true. This project for economic and monetary integration not only falls short from a social, *ecological*, and democratic perspective, but also from an economic one.
>
> The reasoning behind [the Maastricht] convergence criteria is drawn from monetarist doctrines that are not accepted by the majority of economists. According to these doctrines, reduction of budget deficits leads to lower inflation, and lower inflation automatically leads to more growth and employment. Recent economic research by renowned economists... shows that this assertion cannot be verified empirically. (Open Letter from European Economists, 1997; emphasis added)

As has been the case with climate change, scholarly activism that ran counter to the interests of the SCAMD elites was either ridiculed or ignored. Despite warnings by prominent economists, the project of single currency marched on with no hesitation. Quite to the contrary, under the pressures of Maastricht, national governments had to "[subordinate] democracy to the dictates of a neoliberal restructuring of state finances," as was the case in Belgium in 1996, where "a peculiar combination of austerity and absolutism" had to be enforced to meet EMU convergence criteria. "The future," foretold Gill, "may be worse" (Gill, 1998: 17).

Since the outbreak of the euro zone sovereign debt crisis in 2009, many in Europe have experienced first-hand the consequences of the neoliberal art of exclusion through inclusion, as mastered by the global SCAMD elites. Establishing a monetary union based on not sound theoretical and empirical findings but on neoliberal PR imperatives of entrenching the preferred order behind an impassable moat was bound to lead to catastrophes sooner or later. Remaining true to the value that not everyone mattered, the SCAMD elites coped with the financial crisis by crushing and eliminating the unimportant and the irrelevant. The SCAMD elites heavy-handedly administered austerity measures to torn societies even as it was becoming increasingly impossible to hide the fact that the disaster resulted from the doings of these elites in the first place. The creation of the single currency significantly reduced the costs of loans for countries such as Greece, which used to have weak currencies and were faced with higher interest rates when borrowing. Once the euro zone was created, cheaply available loans combined with large banks' predatory lending quickly produced the illusion of economic prosperity in these countries and pushed up unit labor costs relative to countries such as Germany, which have had strong currencies historically. During the decade spanning from the birth of the euro and the outbreak of the euro zone sovereign debt crisis, severe trade imbalances built up between trading partners such as Greece and Germany. The alleviation of the imbalance through currency devaluation, however, was an effective method that no longer existed under the single currency. The continual deterioration led to severe payment imbalances between debtor and creditor countries, further opening up opportunities for predatory financial institutions, until the repayment ability of Greece came under spotlight in 2009 and shook the market. Unsurprisingly, the SCAMD elites had no interests in admitting their role in the "crisis," which left their wealth expanding still further at a remarkable rate.[24]

The SCAMD elites' disregard of human lives is at the core of the foundation of today's socioeconomic order. It makes little sense to differentiate the disregard in the economic realm from that in the environmental realm. The desire to exclude and the necessity to hide such a desire resulted in the neoliberal PR monster. Unlike in Chile, where the SCAMD elites could simply rule by military force, in places such as Europe, appearances of civility were essential, giving rise to sophisticated works for manufacturing consent. While conspiracy theorists of the Bilderberg Group, the birthplace of the EU, may be overexcited, as Kakabadse points out, they do have a point. The way that the Group operates is in fact "much smarter than conspiracy, [since it molds] the way people think so that it seems like there's no alternative to what is happening" (BBC, 2011). If the EU itself is a sophisticated edifice built by the SCAMD elites to advance the neoliberal art of exclusion through inclusion, then it is highly probable that, insofar as climate change is concerned, the EU is part of the problem not only in terms of the slow reaction to the climate crisis, but also in terms of the fermentation of the crisis in the first place.

In the year in which Delors declared that the absence of a single European currency was incompatible with the Single Market, the Intergovernmental Panel on Climate Change (IPCC) published its first major report. In the year that the Treaty of Maastricht was signed, the United Nations Framework Convention on Climate Change (UNFCCC) was adopted during the Rio de Janeiro Earth Summit. In the year that EU member states agreed to the Stability and Growth Pact, the Kyoto Protocol was adopted by the parties to the UNFCCC. It was not in two parallel universes that the SCAMD elites' push for the completion of the single market and the establishment of the single currency on the one hand, and the world's efforts to tackle climate change on the other took place. The force capable of out-maneuvering opposing interests in one realm prevailed in the other precisely because it reigned above both the market and democracy.

Nearly three decades after the first IPCC report, with four further major reports published, annual emissions today are more than 60% higher than in 1990 and are still rising. "Less than 12 years of current emissions will see our 1.5C aspiration go the way of the dodo, with the 2C carbon budget exceeded by the mid-2030s." For Kevin Anderson, "the international community has presided over a quarter of a century of abject failure to deliver any meaningful reduction in absolute global emissions" (2018). Despite the wealth of knowledge accrued on both the problem and what we need to do about it,

> [s]hamefully our responses to the challenges posed by this knowledge have been dominated by a litany of scams: Offsetting[25] (getting the poor to diet for us); CDM[26] (officially sanctioned offsetting); EU ETS[27] (with so many permits issued the CO_2 price remains irrelevant) ... (Anderson, 2017)

The inescapable physics indicates that, for the EU to safeguard any reasonable probability of capping the temperature increase at 2°C, it would need to "make immediate emissions reductions of approximately 10% p.a., arriving at a 2030 decarbonization target of around 80%" compared with 1990" (Anderson, 2013). The actual target that the EU has adopted, instead, is a 40% reduction in greenhouse gas emissions by 2030 (European Council, 2014: 1).[28] Not only has the EU set the target at half of what is actually needed, but it has also declared that the ETS "will be the main European instrument to achieve this target" (European Council, 2014: 2).

The ETS has been the backbone of the EU climate strategy since 2005. It was *not* the instrument that the European Commission had originally advocated in the late 1980s and the beginning of the 1990s, when the ERT's influence in the Commission was still limited to the domain of the single market. Not yet effectively controlled by the SCAMD elites, the Directorate Generals (DGs) responsible for the climate package—DG XI on environment and DG XVII on energy—came very close to pushing through a carbon tax under the leadership of Commissioner Carlo Ripa di Meana (DG XI) (Skjaersethrseth, 1994: 27). In retrospect, the replacement of a carbon tax by carbon trading was predictable, given the influence of the ERT Environment Watchdog Group and the Union of Industrial and Employers' Confederations for Europe (UNICE). As a general rule, proposals of which the SCAMD elites disapproved vanished from the list of potentially feasible policy options. Non-existent measures, in contrast, were born in response to the wishes of elites.

The EC began funding research programs in climatology in 1980. In 1985 the Commission acknowledged that "climate [has] in store a major change which could be brought about by the increase of atmospheric CO_2, mainly due to fossil fuel burning" (European Commission, 1985: 6). In 1986 the European Parliament passed a resolution on counteracting the greenhouse effect (European Parliament, 1986: 272). A month before the IPCC was established in 1988, the Commission presented its work program on options dealing with the greenhouse effect (European Commission, 1988). In March 1990 the Commission issued an outline for policy targets on the climate issue, highlighting the urgent need for industrial countries to commit to stabilizing CO_2 emissions by the year 2000 (Skjaersethrseth, 1994: 26). Climate change was placed on the agenda of the following European Summit for the first time in June 1990. At the summit, member state leaders issued a declaration on "the Environmental Imperative" and pledged to "adopt as soon as possible targets and strategies for limiting emissions of greenhouse gases" and requested the Commission to expedite the drawing up of its action plan with a view to establishing a strong Community position in preparation for the Second World Climate Conference to be held in Rio in June 1992 (European Council, 1990: 27). In October the joint Council of EC Energy and Environment Ministers agreed that, assuming similar commitments from other leading countries, the EC and member states "are willing to take actions aiming at reaching stabilization of ... total CO_2 emissions by 2000 at the 1990 level in the Community as a whole" (Bergesen et al., 1994: 17).[29] To hammer out implementation measures for the target, the Commission started to work on a climate package that involved renewables, energy-efficiency, and a carbon/energy tax. In order to ensure a leadership role for the EC in the global climate negotiations, the Commission aimed for the measures to be adopted before the Rio Earth Summit (Skjaersethrseth, 1994: 25). The goal-formulating process in the period ending in December 1990 went "remarkably rapidly and smoothly," with natural science aspects of the issue embodied in the IPCC conclusions published in August that year "apparently taken for granted." Although the unanimously adopted stabilization target was not legally binding, it carried significant political weight and set off the phase of developing measures aimed at reaching this goal (Skjaersethrseth, 1994: 27).

In February 1991 the Commission began circulating the idea of carbon and energy tax as a key method for achieving the stated target. In a discussion paper drawn up by the environment Commissioner Ripa Di Meana and the energy Commissioner Antonio Cardouso e Cunha, a combined tax and levy on energy and fuels were proposed (Hunt, 1991a). On 23 September 1991 a group of European industrial leaders representing automotive, cement, chemicals, and petroleum industry associations issued a joint statement arguing that the proposed tax "would have severe economic and social consequences and weaken the competitive position of European industries and threaten jobs and prosperity." The group insisted that, instead of a unilateral imposition of such a tax, the Commission should be pushing for concerted international action covering all countries (Hunt, 1991b).

Despite the protest, the Commission agreed on 25 September 1991 to the proposal for an energy tax. Commissioner Ripa di Meana pointed out that while the European Community generated 13% of world CO_2 emissions, compared with 23 % from the U.S., the Community's contribution to global warming was increasing. To meet the target of stabilizing CO_2 emissions by 2000 at the 1990 level, fiscal weapons were the least expensive while sending a signal to both industries and consumers that environmental costs must be internalized (European Commission, 1991: 7). According to the Commission's plan, member states were to introduce the tax incrementally. It would be based on a 50–50 split on the *carbon* content of fossil fuels and on all non-renewable *energy*, and start with a $3 levy in 1993 with a rise each year until 2000 (European Commission, 1991: 8–9). The tax would be offset by tax cuts in other areas, as the goal of introducing the tax was not to increase revenue but to change the balance of the existing tax system in favor of the environment (D. Gardner, 1991a; European Commission, 1991: 7).

The Commission's proposal was welcomed by environment ministers on 1 October 1991. Ripa di Meana described the plan as being warmly received, commenting that "[w]hat has been considered impossible until now is being considered possible" (D. Gardner, 1991a). The *Financial Times* reported "[i]f the threat of global warming is taken even half-seriously there is no alternative to energy taxes." Even where doubts remained, the sensible response was to "adopt the precautionary approach—the dissenting Americans may be right, but the risks inherent in their being wrong are so great that it is prudent to take measures to curb emissions now" (*Financial Times,* 1991). In early October finance ministers gave the plan provisional backing (D. Gardner, 1991b). On 13 December 1991 energy and environment ministers gave the Commission the go-ahead to draw up legislative proposals for the energy tax. At that point, the mood among observers was so optimistic that *The Wall Street Journal* reported that "support is growing within the European Community for an energy tax to combat global warming." Even EC finance ministers, "who up to now have been less enthusiastic than their environment and energy counterparts... are expected to back the call for Brussels to draft legislation." "EC governments increasingly view[ed] taxation as a useful tool in influencing energy consumption," with Spain the only country still hesitating (Wolf, 1991).

The success of the plan in curbing emissions and in enabling global action depended on two contingencies: First, that the proposal would not be severely watered down in its finalization. Second, that the Council would reach a positive decision *before* the Rio Summit, giving Commissioner Ripa di Meana the leverage to pressurize other countries to follow. In the event, "failing miserably" seems the only suitable way to describe the way that the process actually developed. Not only was the tax proposal severely watered down, but even the watered-down version failed to secure the Council's backing before the Rio Summit.

At the stage of finalizing the proposal, it "was made subject to some of the most ferocious lobbying ever seen in Brussels" (Skjaersethrseth, 1994: 28). Already in June 1991 UNICE sent a position paper to the Commission, stating that "such plans run completely counter to the need... for concerted international action" (UNICE, 1991: 155, cited in Skjaersethrseth, 1994: 29). As the proposal continued to move closer to the finish line, UNICE adopted the strategy, in order of priority, of: i) blocking a go-ahead from the Council; and ii) ensuring that the proposal was made conditional on other OECD countries doing the same if the Council approved the Commission's strategy (Skjaersethrseth, 1994: 29).

In drawing up the proposal, the Commission emphasized that while the Community "will have to make every effort to ensure its partners undertake comparable concrete action," the overall strategy proposed by the Commission "can stand on its own and have positive benefits for the Community" (European Commission, 1991: 11). In other words, the Community's action did not have to be conditional on others doing the same. Large industries, however, were not interested in the slightest in the EC being a leader. Resorting to the logic of "I will stop stealing if you stop first," industries once again counted on the magic word of "competitiveness" to justify their insistence on making the EC's action conditional to others doing the same. Initially, Commissioner Ripa di Meana fought hard against the conditionality principle, even threatening to boycott the Rio Summit if he did not get his way (Skjaerseth, 1994: 28). In the end, however, "faced with an alliance between DGs representing tax, internal market interests and the Commission President himself" (Skjaerseth, 1994: 30), and under the criticism that his DG "has often paid more attention to the lobbying of ecological activists than to that of industry" (*The Economist*, 1992a), Ripa di Meana had to give in and accept the principle of conditionality.

On 13 May 1992 the Commission approved its draft directive, which stated that steps must be taken "to avert the risks of European industries being tempted to relocate to third countries where environmental standards are less stringent than in the Community." Concretely, "the Community will be unable to apply the tax until such time as its main competitors within the OECD have introduced a similar tax or measures having an equivalent financial impact" (European Commission, 1992: 4). While it was Ripa di Meana's belief that the Council would approve the proposal in time for him to bring it to Rio that led him to concede to conditionality (Skjaerseth, 1994: 30), green members of the European Parliament and environmental groups condemned the decision of making the EC carbon tax "subject to a veto by the US and Japan" (Palmer, 1992b). Even after Ripa di Meana conceded on the issue of conditionality, his hope that the EC could play the leading role and mobilize global opinion at the Rio Summit was dashed when environment ministers failed to endorse the proposal at their 26 May meeting (Wolf, 1992). This development showed that the alleged concern over competitive disadvantage, a problem solved by conditionality, was just an excuse for blocking climate mitigation measures that would undermine the interests of the industries. Industries argued that a tax, even a conditional one, was unnecessary, as firms were already voluntarily becoming more energy efficient (*The Economist*, 1992b: 85).

It is not difficult to see that behind the colossal double failure of the Commission's attempt to tax carbon and to lead the world in climate mitigation, the SCAMD elite power was increasingly able to prevent global actions of which they disapproved. Infrastructure was one of six key focal points that the ERT had identified at its foundation. In 1984 the ERT published a document called *Missing Links* and in 1991 *Missing Networks*, sketching out the ambition of turning Europe's infrastructure into "a single interacting system of mega-network with a single output: mobility" (Balanyá et al., 2003: 22). Insisting that existing infrastructure was inadequate and a barrier to unhindered economic growth, the ERT pushed the Commission to adopt the environmentally controversial Trans-European Networks (TENs) (Balanyá et al., 2003: 22–3).[30] Greenpeace estimated that the construction of TENs would lead to an increase of approximately 15–18% of greenhouse gas emissions from the transport sector (Balanyá et al., 2003: 69). Despite such concerns, the ERT had the TENs placed squarely on the EU's agenda (Balanyá et al., 2003: 23). Volvo CEO Pehr Gyllenhammar boasted that the transport ministers of the EC were "referring to *Missing Links* as the masterplan for European infrastructure" (Balanyá et al., 2003: 69).

Besides the strong influence from the ERT, two of the three Commissioners who adamantly opposed the tax proposal in the 17-member college were participants of Bilderberg conferences. Martin Bangemann, the industrial affairs Commissioner, was a Bilderberg attendee in 1986. Sir Leon Brittan, Commissioner for Competition Policy, participated in the 1992 Bilderberg conference on 21–24 May, between the crucial Commission meeting of 29 April, where heated debate on the carbon tax lasted over six hours, and the crucial Council meeting of 26 May that dashed Commissioner Ripa di Meana's hopes of an EC commitment before Rio. Together with the Commissioner on Taxation Christian Scrivener, they called for the plan to be delayed and for the proposal to be downgraded from draft directive to working paper (Palmer, 1992a). Commissioner Scrivener had been the leading opponent of the tax plan from the beginning. She was superb at the neoliberal language: "The environment must not be used as a pretext for an increase in the overall tax burden in Europe." "European industry must not be penalized." "It is just impossible to say a tax will solve everything." She emphasized that "[t]he introduction of any new taxes Europe-wide in the run-up to the single market in 1993 would be 'disastrous' " (Jack, 1991).

As the mantra of "competitive disadvantage" was repeated in the campaign against the Commission's proposal, opponents of the plan sat harmoniously side by side with their foreign trade "competitors," producing the business manifesto for the Rio Earth summit (Cowe, 1992).[31] While Ripa di Meana was busy defending his tax proposal in Brussels, the British government took the initiative to weaken the world treaty on the greenhouse effect, which was about to be signed in Rio. In the meantime, Chancellor Helmut Kohl of Germany, Prime Minister Lubbers of the Netherlands, and leaders of other countries received calls from U.S. President George Bush, Sr, asking for reassurance that there would be no firm target in the treaty of stabilizing carbon emissions at 1990 levels by 2000 (Lean and Ghazi, 1992).

Instead of going to Rio empty-handed, Commissioner Ripa di Meana stayed in Brussels and published an op-ed entitled "Why I'm not in Rio." He explained that the failure of EC member states to endorse his plan before the Rio summit and the voiding of all content of the world treaty on the greenhouse effect had prompted his decision not to attend: "I do not believe that there will be any real negotiations… It is my feeling that everything has been 'stitched up' in advance and that any changes adopted will be of the cosmetic rather than the substantive variety" (Ripa di Meana, 1992).

Whichever way one looks at it, within global society, the cleavage was *not* between Europe and the U.S., Japan, or other countries as the "competitive disadvantage" narratives suggested, but between those defending a neoliberal exclusionary grasping order and those that stood to lose in such an order. *The Economist* was frank about the changing face of social cleavages: "The rise of nation states produced national ruling classes. It would be odd if the current integration of the world economy did not produce new global elites," what David Rothkopf called a "global superclass." In this global superclass, global companies are run by business people and financiers, and supranational organizations such as the EU and the IMF are steered by global politicians. A central feature of this structure is that "[t]hey operate on the global stage, far from mere national electorates" (*The Economist*, 2008). By the time of the Rio Summit, international trade was increasingly an intra-corporate affair: over 40% of it was carried out *within* transnational corporations. As a result, the top 500 companies in the world were controlling about 70% of world trade,

80% of foreign investment, and 30% of world GDP. These 500 companies generated more than half of the total greenhouse emissions produced by global industry (Watkins, 1992; Vidal, 1992).

Hutton used the term "green Keynesians" to capture the thinking of environmentalists fighting at the Rio Summit. Just as Keynes did not trust to leave everything to the market—particularly because "in the long run, we are all dead," the environmentalists did not trust that the natural system could return, in time, to balance, once it started interacting with the modern growth-oriented industrial economies. The significant role of government in regulating what would otherwise be destructive was therefore justified (Hutton, 1992). In contrast, Fred Smith, the former head of the anti-environmental think tank Competitive Enterprise Institute, ridiculed those who wanted to bring in "meddlesome government agency, international agreement or market-distorting tax," asking "who is to say that global warming will not mean that the Earth provides its own self-correcting mechanisms, such as more clouds?" (Hutton, 1992)

Within the EC, lobbyists and corporate consultants were flocking to Brussels to ensure their favored treatment in the Single Market. A partner of a law and government relations firm in the U.S. offered an insider's peek into the dramatic change taking place in Brussels, a city "brimming with lobbyists." Large U.S. law firms with strong Washington, DC lobbying experience led the way to join

> an international armada of advocates already active in Brussels, including... public relations groups, confederations of European trade associations, representatives of US states, German lander and British municipalities, small 'boutique' consultancies, in-house representatives of individual US, European, and Japanese companies, European trade unions, agricultural groups, and a growing number of public-interest associations. (J. Gardner, 1991a: 29–30)

"Brussels," James Gardner wrote, "is very much an insider's town." Within the "growing swarm of lobbyists," some enjoyed better access, such as the EC Committee of the American Chamber of Commerce in Belgium (J. Gardner, 1991b: 39). The influence of the Committee over EC legislation was "out of proportion to its membership census" (J. Gardner, 1991b: 42). The Committee's remarkable success was attributable to the hybrid form of lobbying which "tempers the best features of Washington, DC lobbying—clear, unambiguous, fact-based presentations—with the low-key, accommodating style favored by Europeans." With a triumphant tone, Gardner noted that the Committee's effectiveness was "indisputable proof that foreign interests can lobby the institutions of the European Community successfully."

> As one member of the committee said, "We keep hearing complaints from US Department of Commerce officials and the US Trade

Representative that the US doesn't have a "seat at the table" in the EC '92 process." The truth is that the EC Committee has had a seat at the table from day one. (J. Gardner, 1991b: 43–4)

Given the weight of the Committee, it is not surprising that the group had attracted "the financial support of the European corporate community, including a handful of European firms of American parentage like Merck, Monsanto Europe, 3M Europe, DuPont, Exxon Chemical and Honeywell Europe" (J. Gardner, 1991b: 50). By 1994 lobbying had become one of the fastest-growing industries in Brussels. With 3,000 groups employing around 10,000 lobbyists, the ratio of lobbyist to commission officials was roughly 1:1.3, double the level in 1990. By 1994 it was no longer necessary to speak about lobbying in Brussels in a hushed tone. The contrast of attitudes between Europe and the U.S. was almost reversed. A U.S.-based lobbyist specializing in "Eurolobbying" wrote in an upbeat tone that, "[w]hile lobbying can carry negative baggage in Washington, things are more positive in Brussels." The Commission, for instance, "openly and actively helps companies and associations with their lobbying by freely supplying information and technical advice necessary for lobbyists to make their arguments." In return, lobbyists helped the Commission to see the big picture: "The American Chamber of Commerce has been recognized for its work in preparing detailed and articulate reports on various issues" (Facchinetti, 1994). The deputy head of Commissioner Scrivener's cabinet, Michel Petite, explained that lobbying in Brussels was a "sophisticated, formal process that focuses on substance and involves careful documentation" because "busy *bureaucrats appreciate* research support and *drafting assistance*" (J. Gardner, 1991b: 56; emphasis added).

With Ripa di Meana resigning in protest and the flagship carbon tax proposal shelved, the Community in effect left climate policy to the member states to elaborate national emissions reduction programs. At the negotiations of the 1997 Kyoto Protocol, developed nations agreed to decrease emissions by at least 5% below 1990 levels by 2012. Disagreements over what methods should be applied to meet the target brought the negotiations to the brink of collapse. The U.S. under the Bill Clinton administration wanted to create an international market for carbon emissions for polluters to purchase and sell their allowances. In contrast, the EU's proclaimed favored method was carbon tax. "Radically antagonistic" was how France's then environment minister, Dominique Voynet, described the conflicts between the US and the EU. For her, to rely on the market to solve the climate crisis was to abandon the crisis to "the law of the jungle." Similarly, Germany's environment minister criticized the US proposal saying that "[t]he aim cannot be for industrialized countries to satisfy their obligations solely through emissions trading and profit" (Klein, 2014: 218). The conflict ended with the U.S. winning the battle at the negotiating table only to abandon the Protocol altogether three and a half years later. While Naomi

Klein described the US pulling out of the Kyoto Protocol and the EU embracing emissions trading as "one of the great ironies of environmental history" (Klein, 2014: 219), the EU might not have been as enthusiastic about the carbon tax as it tried to appear. Given the necessity, in the neoliberal art of exclusion through inclusion, of allocating different roles (good-cop/bad-cop) to U.S. and European players, labelling carbon trading as "American" might have been an over-simplification.

The ERT Environment Watchdog Group published a working paper on climate change in 1994. The paper considered "scientific uncertainty" and "global dimension" as the most important features of the challenge. With respect to the former, it emphasized that scientists were "grappling with incomplete data," as "even the most advanced state-of-the-art computers cannot handle all the input data and feedbacks required to predict climate trends with accuracy on a global scale" (ERT, 1994: 8–9). With respect to the latter, it emphasized that "climate change is a global problem which cannot be solved by actions taken in isolation.... The remedies will only be workable if they are agreed internationally." As a result, the group considered it essential that "measures used... do not damage the international competitiveness of EU industry" (ERT, 1994: 8). For the way forward, the paper was completely dismissive of "command-and-control" regulations (ERT, 1994: 17). It then criticized the EU's carbon/energy tax proposal on the grounds that it not only ignored voluntary initiatives from industry but also failed to be cost effective economically. As a result, the tax proposal had "little relation to the objective of limiting CO_2 emissions" and would be "ineffective, distortive, and damaging to industry's competitiveness" (ERT, 1994: 21). To "motivate and aid energy users to lower their CO_2 emissions levels," industry "is ready to sit at the table with government officials to devise more effective economic instruments" which included "tradeable permits" (ERT, 1994: 19; 21). In 1997 on the eve of the Kyoto meeting, the ERT Environment Watchdog Group published another report on climate change. After re-emphasizing scientific uncertainty, the global dimension of the problem, and the importance of European competitiveness, the report emphasized the need for policymakers to consult with stakeholders. The cover of the report said that "Governments intend to take action to reduce greenhouse gas emissions." "Industry **must be part** of the process" (original emphasis). To make it easy for policymakers to understand, the report drew up lists for "how to get it right" and "how to get it wrong." Numerous criteria were listed for the first category. As to the list for "how to get it wrong," only one thing was mentioned, namely, the EU's carbon/energy tax proposals (ERT, 1997: 7). With explicit support from the U.S. and covert support from Europe, voluntary emissions trading was written into the Kyoto Protocol as an option for emissions reduction. The headline of a *Guardian* report captured the essence of voluntary emissions trading: "For Sale: The Right Not to Use This Factory" (Brown and Cowe, 1997).

As the approach most favored by neoliberals, emissions trading does have the problem of betraying one of their darkest secrets. Polanyi did not believe that the market was self-regulating. Keynes thought that there were times when the self-regulating function of the market was irrelevant (as in the long run, we are all dead), rendering strong government actions not only justifiable but desirable. Neoliberalism defeated both by selling the doctrine that the market is self-regulating and that any intervention on the part of the government can only make things worse. Emissions trading is a complete artificial market.[32] It has nothing to do with two parties having a mutually beneficial demand-supply relationship spontaneously finding each other in the marketplace. Quite the contrary: the setting-up of an emissions market requires the government to play the "self-regulating" role of the market— from determining the "cap," to defining the commodity, to price-setting— arbitrarily. The claim that emissions trading could work gave the neoliberal game away and showed neoliberals' true belief that the market had never been self-regulating. This reveals that neoliberals are Keynesian by heart, but staunchly anti-Keynesian in words and deeds for the sake of exclusion.

After the Kyoto meeting, the EU waited six months before publicly aligning its position with that mandated by the SCAMD elites. As if the Protocol had pointed to a gem, the beauty of which the Commission had previously failed to appreciate, the Commission's Communication on the EU's post-Kyoto strategy used "flexible" or "flexibility" 53 times and "trading" in the context of "emissions trading" 80 times, in contrast to the mentioning of "tax" or "taxation" just nine times. The Commission declared the imperative of building an *EC-wide* emissions permit market system in order to prevent discrimination and the distortion of competitiveness. In view of the lack of experience, the Commission recommended a prudent step-by-step approach, with the goal of setting up the internal trading regime by 2005 (European Commission, 1998: 19–20). The problem of a lack of experience was swiftly solved by inviting a Washington-based think tank, the Center for Clean Air Policy (CCAP), to help to design the system. Founded in 1985, the Center helps policymakers to "develop, promote and implement innovative, *market-based* solutions to major climate, air quality and energy problems that *balance both environmental and economic interests*." Its mission is to "significantly advance *cost-effective* and *pragmatic* [policies] through analysis, dialogue and education to reach a broad range of policy-makers and *stakeholders worldwide*" (CCAP website, emphasis added). Its website notes that its funding comes from private foundations, U.S. federal, state, local, and international governments, individual donors and corporations. It received grants from Exxon Mobil in 2007, 2009, and 2014.[33] The Center submitted its final report on EC emissions trading to the Commission in 1999 (CCAP, 1999), which was clearly concerned only with cost effectiveness and stakeholder benefits and not with the effects of the proposed scheme on emissions reductions. The CCAP was not the only organization from which the Commission sought help. The

Foundation for International Environmental Law and Development (FIELD)[34] was also commissioned to work on the design of the emissions trading regime. In the chapter entitled "The Case for Emissions Trading in Europe," the team emphasized that emissions trading was better than "command-and-control instruments, however smartly designed," because "[t]he process of cost-minimization, left to the market, rather than the regulator, is endogenous and adaptive." As a result, it was "a corrective, rather than a distortionary measure... [for tackling] a global environmental concern caused by market failure" (FIELD, 2000: 17).

BP and Shell also played an important role in the formation and adoption of the EU emissions trading scheme (ETS). By 2000 both companies had already installed company-wide internal emissions trading systems. The rationale was to use shadow pricing[35] to curtail the financial risk of stranding assets, which might result from imposing carbon prices on polluters following the Kyoto agreement. In the process, BP found that an over-optimistic calculation of business growth would lead to over-allocation of allowances, which would mean that the pledged target of emissions reductions would easily be met. This had a significant impact on the way rules of the EU ETS were set, with industries insisting on a free handout of credits from the outset (Comello and Reichelstein, 2014:14; Gilbertson and Reyes, 2009: 29). Shell's 2002 report stated "we welcome the EU proposals for a mandatory, EU-wide emissions trading scheme. We have completed a three-year internal CO_2 trading trial and are sharing our knowledge and experience with governments." The report quoted the key figure in making emissions trading the main policy instrument for EU's climate policy, Jos Delbeke, Director of Environment DG in the Commission, saying:

> Shell, as a world leader in the energy business, is an example to be copied insofar as it writes climate change into its business plan. As a result of Shell's own work in developing GHG emissions trading, and as befits a 'first-mover,' Shell will be better prepared than most when the EU's emissions trading scheme starts. (Delbeke, quoted in Shell, 2002: 28)

Based on the reports of CCAP, the FIELD and input from BP and Shell, the Commission published the green paper on emissions trading in March 2000 (European Commission, 2000). In 2003 the EU adopted a directive for the world's first-ever international cap and trade system. In 2005 the EU ETS started operating. The ETS introduced a brand new fictitious commodity for buyers and sellers. At COP 6 at The Hague, just a few months after the Commission published its 2000 green paper on emissions trading, the conference center was "teeming with lobbyists from emissions brokerage firms, many from global consultancy giants like PricewaterhouseCoopers and banks that have discovered 'the carbon economy'." The emissions brokers were united in lobby groups, the biggest being the Emissions Marketing Association that brought 48 lobbyists to the conference. Specialized magazines such as the *Carbon Trader* and the *Carbon Market Analyst* "flooded the conference center with expensive glossy brochures," promising that "the market in carbon emission credits could grow to trillions of US dollars over the next decades" (Ma'anit, 2000).

The colonization of the UN climate talks by "market mania" turned such events into something "more [like] a trade fair than an intergovernmental conference looking at ways to solve one of the world's most pressing environmental and social problems"[36] (Ma'anit, 2000). The concept of "polluter pays" became quite absurd in the atmosphere. *The Economist*, for instance, blamed the continuing delay in turning the Kyoto targets into a workable program on the tendency of environmentalists to "cast the issue as a *moral* question rather than as an *economic* one." The mindset that pollution is sin, meaning that polluters must be punished, "has infected many governments," contributing to their failure to feel satisfied even when the U.S. came up with ways to reduce global emissions that "do not inconvenience it very much—such as investing in carbon 'sinks' or buying emissions permits from other countries" (*The Economist*, 2000: 78). Within this context, it seems that the green paper's explicit mention that "[i]t is the Commission's view that the involvement of *companies* [as trading entities] in emissions trading represents a unique opportunity for a cost-effective implementation of the Kyoto commitments" (European Commission, 2000: 9; emphasis added) was well synchronized with the actions of the teeming emissions brokers and the clients that they represented.

Doing what it was designed to do, namely, "awarding profits to polluters and encouraging continued investment in fossil fuel-based technologies while disadvantaging industry focused on transition away from fossil fuels," the ETS is costing the world dearly (Gilbertson and Reyes, 2009: 31). For each year of its operation, the EU ETS has continued to "enclose and privatize the global atmospheric commons—awarding property rights to polluting companies" (Gilbertson and Reyes 2009: 31). The ETS was designed in such a way that the targets can generally be met without actual ETS-instigated reductions taking place. The website of the EU Climate Action states that for phase 1 (2005–07) of the ETS, "the total amount of allowances issued exceeded emissions... with supply significantly exceeding demand." For phase 2 (2008–12), the cap went 6.5% lower compared with 2005. During phase 3 (2013–20), the cap decreases each year by a linear reduction factor of 1.7% of the average issued annually in 2008–12. In plain language, the EU has continued to grant the pollution rights to private firms more generously than the polluters have needed to cover their existing level of emissions. Surplus allowances are then sold to other polluters for them to avoid emissions cut as well. Apart from capping the emissions at too high a level in general, the ETS also distributes emissions allowances in a way that rewards big polluters and punishes environmentally conscious small and medium-sized businesses. The Commission pointed out in the green paper that there were two methods that could be combined in the scheme: auctioning and allocation free of charge. "In the context of emissions trading, the latter is often referred to as 'grandfathering'." The Commission explained in a footnote that, strictly speaking, a grandfathered right was "not related to the notion of the allocation free of charge of a realizable asset, but rather to a historical right to do something, such as vote, that can be transmitted to descendants or retained by a legal entity" (European Commission, 2000: 18). A whopping 95% of all allowances allocated to industries had to be grandfathered for phase 1 of the ETS and 90% for phase 2. What the ETS allocation methods essentially did, therefore, was to bestow the quality of property right to the right to pollute. Emissions allowances are registered as an "asset" on companies' books, the maximization of whose values is the obligation of the management team. Small and medium-sized businesses fail to qualify as recipients of such cost-free asset rewards because they are not major polluters. Debelke confirmed that the logic of free allocation based on historical emissions was indeed one of "the more one pollutes, the more allowances one gets..., with the negative effect of favoring less efficient facilities" (Gilbertson and Reyes, 2009: 33).

The effective upward redistribution of the ETS is not confined to the EU itself. By treating historic pollution as an earned credential licensing corporations for more future pollution, the scheme extends such rights globally by linking the ETS to the largest global "carbon offset" program, the UN Clean Development Mechanism (CDM). Carbon offset is a system that allows polluters to avoid their duty to cut emissions by paying money to allegedly improve someone else's emissions performance. Airlines, for instance, can proudly announce that passengers no longer need to feel guilty about flying because they can now offset the carbon emitted from the trip by paying for the airlines' projects to plant trees, provide long-life light bulbs, or construct hydroelectric dams in poor countries (Davies, 2007). As the former chief economist of the World Bank, Larry Summers, infamously remarked, "the economic logic of dumping a load of toxic waste in the lowest wage country is impeccable," and that "under-populated countries in Africa are vastly under-polluted" (Gilbertson and Reyes, 2009: 27).

As there is no reliable way to verify the claimed effects, compared with what would have happened in the absence of the light bulbs or hydroelectric dams, Welch describes offsets as "an imaginary commodity created by deducing what you hope happens from what you guess would have happened" (2007). Even when the problem of verification is placed aside, the notion that developed countries—the major historic polluters—should be entitled to continue polluting the atmosphere in their insatiable pursuit of economic growth so long as they pay some poor countries to stop polluting can be described as nothing short of colonialism. At the Rio Summit, developed countries used obscure wording and inserted in the Treaty a clause stating "Parties may implement [mitigation] policies and measures *jointly* with other Parties" (UNFCCC, Article 4.2 (a)). With a sickening turn, the concept of "joint implementation" that allowed rich countries to shirk responsibilities further contributed to the creation of the CDM, which enabled large polluters to not only avoid responsibilities but also make money by doing it. At Kyoto, concerned with the effect of joint implementation on exempting responsibilities of developed countries, the Brazilian government proposed the establishment of a Clean Development Fund to neutralize the exempting effect. Under the proposed plan, fines paid by developed countries exceeding their targets would be used to finance clean energy in poor countries. Rich countries took this idea and transformed it into the CDM, turning the claim of helping poor countries into credits that can be purchased. Instead of a fund, a trading mechanism was established, with fines transformed into prices and a judicial system into a market (Gilbertson and Reyes, 2009: 27–8). The creation of the EU ETS further helped traders, consultants, large financial players, and big polluters to tap into the CDM treasure even more easily.

Neoliberals would be quick to point out that, despite all its flaws, the EU ETS does work, as in late 2018 the price of carbon finally shot up to €24.85 a ton, placing pressure on coal. Climate change would not be a crisis, were time not the most menacing dimension of the problem. The World Bank's High-Level Commission on Carbon Prices, chaired by Joseph Stiglitz and Nicholas Stern, published a report in May 2017, indicating that a carbon price of at least \$40–80 (€35–70)/ton CO_2 by 2020 and \$50–100 (€45–90)/ton CO_2 by 2030 would be necessary to meet the Paris Agreement goal of 2°C (World Bank, 2017). The rise of carbon prices to a fraction of the desired level, two decades after the initial decision was taken to make cap and trade the flagship strategy for emissions reduction, is no cause for celebration. Had the amount of time needed for the strategy to begin to show signs of effectiveness been part of an open and honest discussion in the late 1990s, cap and trade would likely have been ruled out as a plausible option. The leap of carbon prices in 2018 was partly the result of ETS reforms, introduced by the EU in February 2018. The reform itself contradicts the EU's own claim that the ETS has been working as intended. Year after year, the European Commission claimed that the ETS functioned well. The 2017 annual report, for instance, read that, "[i]n 2016, the EU ETS... delivered emission reductions of 2.9%" (European Commission, 2017a: 37). Similarly, its 2016 report stated that "[i]n 2015 the EU ETS has delivered further emissions reductions... by around 0.4% [which] confirms the decreasing trend over the last five years" (European Commission, 2017b: 34). Reports from previous years included more of the same.[37]

Yet the system that the Commission insists has been working has had to undergo successive reforms, which "repeatedly brought [it back] from the dead," with the latest one ensuring the existence of the scheme to at least 2030 (Reyes and Balanyá, 2016: 4). Before the finalization of the latest reform, big polluters lobbied heavily on the issue of "carbon leakage." Threatening (yet again) to invest elsewhere with more flexible environmental rules if the ETS became too much of a burden, they argued that such a scenario would result in higher overall emissions that nobody wants to see. While even the most thorough study of the issue, funded by the Commission itself, found "no evidence for any carbon leakage" in accordance with other similar studies (Reyes and Balanyá, 2016: 6), the revised ETS directive contains new provisions for protecting industries against the risk of carbon leakage.[38] Apart from the safeguard against carbon leakage, the reform also had to add new mechanisms such as back-loading and the Market Stability Reserve to soak up glutted emissions. These are mechanisms that need not have existed, had the system worked as the EU insisted that it did.[39]

Against the criticism that a system with constant over-allocation of allowances cannot work, the EU argues that over-allocation is a reaction to economic crisis, an external factor, which has reduced emissions more than anticipated. Not only does this contradict the fact that the price of allowances fell to zero in 2007 *before* the economic crisis took place, but it also contradicts the EU's attribution of emission abatement in the post-crisis period to the ETS. Even the Commission's own *ex-post* evaluation indicates that the ETS abatement has a quite limited role in CO_2 emissions reduction, particularly compared with renewables abatement, which started to play an increasingly significant role long before the establishment of the ETS (European Commission, 2016: 15–6). This conclusion is consistent with other studies demonstrating that emissions abatement can be explained almost entirely by the economic crisis, the switching of fuel from coal to gas, increases in renewable energy, and improved energy efficiency (Corporate Europe Observatory, 2015). In addition, as Peters et al. (2011) point out, many of the EU's emissions "reductions" were merely emission "transfers" that resulted from international trade. For the period between 1990 and 2008, for instance, the EU's net emission transfers to countries such as China, India, and Brazil far exceeded the EU's net reduction in territorial emissions (Peters et al., 2011: 8905–6). Moreover, the cause of large emissions transfers in both the US and the EU were likely "preexisting policies and socioeconomic factors" that were "unrelated to climate policy," such as the ETS (Peters et al., 2011: 8907). In a similar vein, using actual historic data as opposed to artificially created hypothetical emissions as the basis for *ex-post* comparison, Bel and Joseph found that, by far, the biggest share of emissions abatement in the EU was attributable to the economic crisis (2015: 538). Likewise, through an *ex-ante* study of EU emissions for the period of 2013–30, Hu et al. found that the ETS will not start to have an impact on emissions before 2025, owing to the prevalence of a sizable allowance surplus, and neither back-loading nor market stability reserve could succeed in restoring scarcity of allowances (Hu et al., 2015: 152).

In a neoliberalized world, countless amounts of money, time, energy, and talent have been and still are directed towards engaging debates like these, when the nature of the problem is not about science but the asymmetrical power distribution under the SCAMD structure. The neoliberals would very much like for such debates to go on forever and remain far away from the core problem. The ETS not only fails to cut emissions and buys time for big polluters, but also crowds out other more effective tools such as carbon taxes and government regulations, making it an "anti-climate policy" (Morris, 2013: 10–11). One of the most striking examples of how the ETS serves to undermine existing environmental regulations is the amendment of the EU's legislation on the IPPC (International Pollution Prevention and Control) Directive. The amendment excludes CO_2 from emissions limits set by the IPPC Directive, with the rationale that "operators... might be obliged to reduce their emissions in order to comply with the IPPC Directive when it could be more economically efficient to increase emissions further and buy additional CO2 allowances instead" (Reyes, 2014: 4; Gilbertson and Reyes, 2009: 21). Likewise, the British government sought to weaken energy efficiency measures and renewable energy targets to prevent the carbon price from collapsing. In the same vein, advisers to the EU's DG on Climate Action warned against tough efficiency measures on the grounds that the carbon price could collapse to zero (Reyes, 2014: 4).

The aspects of the ETS that have performed marvelously are those that have facilitated the upward redistribution of wealth. The ETS offers major polluters brand new ways to make money. Lobbying for excessive free allowances when the carbon price is high has been a major driving force for over-allocation. According to a study by CE Delft, the ETS emissions allowances allocated between 2008 and 2012 were almost 30% higher than the verified emissions for the EU countries that are also OECD members (CE Delft, 2016: 14). Passing on to customers the "cost" of allowances as if they were not free is another effective way to make money (Laing et al., 2014). Together with using cheap credits obtained abroad (e.g., through the CDM) for emissions compliance and save free allowances for sale later, big polluters have made windfall profits that translate into further lobbying power. A study showed that between 2008 and 2015 major European polluters have made over €25 billion in windfall profits from the ETS, with approximately €7.5 billion from selling unneeded free allowances, €16.8 billion from cost pass-through, and €0.8 billion from the price differences between domestic and international credits (Carbon Market Watch, 2016: 3-4).[40] It is not difficult to see why ETS reform has been compared with "red meat for the hungry pack of lobbyists that work the corridors of Brussels' political institutions"(Reyes and Balanyá, 2016: 4).

Notes

1 The cumulative CO_2 emissions of the EU 28 up to 2012 was 357Gt, taking up 18% of world cumulative emissions (Rocha et al., 2015: 8). The cumulative CO_2 emissions of the EU 25 up to 2002 was 26.5% of world cumulative emissions (Baumert et al., 2005: 32). The corresponding shares for the U.S. in these periods were 29.3% and 22%.

2 Apart from Prince Bernhard who sat on the boards of Royal Dutch-Shell, prominent oil figures included Gerrit Wagner, who ran the seven-man committee of Royal Dutch Shell (*Time*, 1972), Emilio Gabriel Collado, who joined the Exxon Corporation in 1947, later became the director of the company and served on the Bilderberg Steering committee (Lewis, 1995), Denis Greenhill of BP, who gave a paper at the 1974 Bilderberg meeting (Campbell, 2000, see also Wikispooks under "Denis Grenhill"), Robert O. Anderson, whom *The New York Times* described as "an oilman whose Stetson-size accomplishments included building Atlantic Richfield into an industry giant [and] discovering oil in Alaska," (Martin, 2007), and George C. McGhee, who served at the U.S. State Department and later became a senior executive of Mobil Oil (Engdahl, 2004: 135). Present at the fateful 1973 Bilderberg meeting discussed in the previous chapter were "the CEOs of Royal Dutch Shell, British Petroleum, Total S.A., ENI, Exxon, as well as significant banking interests and individuals such as Baron Edmond de Rothschild and David Rockefeller, and... Henry Kissinger" (Marshall, 2009). The Report of the Group's 1974 meeting in Megive, France stated that they "managed in effect to set the world's monetary system working again" after the oil crisis (Bertell, 1996). For the 2018 Bilderberg meeting, Ben van Beurden of Royal Dutch Shell and Brian Gilvary of BP, were among the attendees.

3 George McGhee also stated that "I believe you could say the Treaty of Rome which brought the Common Market into being, was nurtured at these meetings" (Thompson, 1980: 170). See also Wilford (2003: 70) and Pieczewski (2010: 581). Pieczewski stated "[a]lmost every father of the Treaties of Rome and Paris was either a member of the European Movement or a participant of the Congress of Europe held in The Hague, or else would be invited to meetings held by the Bilderberg Group" (2010: 598). Similarly, "[o]fficials of the European Community—normally the president goes—have attended all Bilderberg meetings, [including] Walter Hallstein, first EEC president" (1980: 170).

4 According to Aubourg, the ELEC was the breeding ground for the Bilderberg group. It offered a pattern for the Bilderberg group to copy from: "discrete discussions in small circles, relying on personal contacts among elites, generally free-trade oriented. Many of the ELEC members were contacted in the early years to attend Bilderberg conferences" (Aubourg, 2003: 93). See also Wilford, 2003: 79–80.

5 Currently the President of Corporate Social Responsibility Europe (CSR Europe).

6 However, he added that it is purely common sense that business influences society and politics influences society.

7 Before founding Bilderberg, Retinger had already been a key figure in advocating European integration. According to Beddington-Behrens, Retinger was the one "who inspired the creation of the European Movement which brought about the Council of Europe. The whole development of the idea of the unity of Europe, the creation of the Common Market and E.F.T.A. are but some of its political consequences" (1960).

8 https://file.wikileaks.org/file/bilderberg-meetings-report-1955.pdf

9 According to Pieczewski, "Retinger would often quote the words of his companion Paul van Zeeland: 'Europe will have to create a common currency, otherwise there won't be any united Europe'" (2010: 588).

10 This is not to deny that preventing war was the main original goal of European integration. Such a higher goal, however, was not immune to corruption and manipulation by elites, who helped to shape the integration project in such a way so that greed, exclusion, and elite acquisition could not only be accommodated

but also facilitated. I am indebted to Bernice Maxton-Lee for reminding me not to use too broad a brush in portraying European integration.

11 Davignon was appointed in November 1974 as the chairman of the governing board of the International Energy Agency created to "tackle Oil Crisis," (See *Encyclopedia of World Biography*, under "Viscount Davignon") his relationship with Big Oil runs deep. According to Carroll and Carson, Davignon is one of the six most powerful persons forming the nucleus of global corporate network (Carroll and Carson, 2003: 82–4).

12 Etienne Davignon and Pehr Gyllenhammar both served as director at Henry Kissinger's consulting firm. Among its important clients was Umberto Agnelli of Fiat (Nollert and Fielder, 2000: 189).

13 Umberto Agnelli, Fiat, Italy; Sir Peter Baxendell, Shell, United Kingdom; Carlo de Benedetti, Olivetti, Italy; Wisse Dekker, Philips, the Netherlands; Kenneth Durham, Unilever, United Kingdom; Roger Faroux, Saint-Gobain, France; Pehr Gyllenhammar, Volvo, Sweden; Bernard Hanon, Renault, France; John Harvey-Jones, Imperial Chemical Industries, United Kingdom; Olivier Lecerf, Lafarge Coppeé, France; Helmut Maucher, Nestlé, Switzerland; Hans Merkle, Bosch, Germany; Curt Nicolin, Asea, Sweden; Louis von Planta, Ciba-Geigy, Switzerland; Antoine Riboud, BSN, France; Wolfgang Seelig, Siemens, Germany; Dieter Spethmann, Thyssen AG, Germany (Cowles, 1995: 506).

14 For Giovanni Agnelli's involvement in Bilderberg's Steering Committee, see Aubourg, 2003: 96. See also "Bilderberg and the Agnellis" at bilderbergmeetings.co.uk/2018/agnellis/.

15 For extensive discussion of the overlap between the Bilderberg Group and the ERT see Nollert and Fielder (2000: 190).

16 Philips Industries is a frequenter of Bilderberg Conferences. H.F. van Walsem of Philips was at the 1954 (and later 1956) founding Bilderberg Conference. Wisse Dekker attended Bilderberg Conferences in 1982 and 1984.

17 According to former Commissioner Peter Sutherland, "one can argue that the whole completion of the internal market project was initiated not by governments but by the Round Table" (Van Apeldoorn, 2019: 130).

18 Padoa-Schioppa, 1987.

19 *The Economist* magazine called Rewe Zentral AG, the plaintiff in the case, "an unsung European hero," whose contribution to the single market should be "toasted regularly in kir." The company brought the regulatory agency of West Germany that banned the importation to the country of the liqueur, Cassis de Dijon, to the European Court of Justice. The magazine foretold that the Cassis ruling was to usher in *competitive lawmaking* in European countries, as companies and people would be allowed to vote for the member state that "offers them laws with *the right blend of freedom and responsibility.*" No need to worry about a scenario where the prevalence of freedom erodes responsibility, reassured the magazine, as "America's internal market... shows that '*competitive rulemaking*' has not led to anarchy there" (*The Economist*, 1988: 11, emphasis added).

20 Sandholtz and Zysman considered, for instance, the 1970s to be the era of Europessimism: "Europe seemed unable to adjust to the changed circumstances of international growth and competition after the oil shock." In the U.S., on the other hand, "flexibility of the labor market—meaning the ability to fire workers and reduce real wages—seemed to assure jobs... despite a deteriorating industrial position in global markets" (1989: 109–10).

21 The term was coined by German economist Herbert Giersch, a participant in 1975 Bilderberg Conference and president of the Mont Pelerin Society from 1986 to 1988 (Giersch, 1985).

22 Ziltener compared the effects the Commission and the ERT had claimed the single market would create for European economies with actual empirical data in 2004. Despite "the theoretical models of integration and econometric simulations that have been at the forefront of scholarly debate over the last fifteen years," Ziltener found "no convincing empirical evidence of European integration having led to either short-term or sustained economic growth effects." In contrast, he found that "[t]he regulatory changes of the Single Market project were part of a global process of economic restructuring and mainly served to enhance the competitiveness of world-market oriented European companies" (2004: 953). Robert Mundell also estimated the welfare effects from customs reduction to be almost negligible. Ziltener talked about the Commission's trick of "hybrid simulations and big numbers" where in its attempt to produce a big number, the Commission "must have decided whenever there was a choice that the correct number was the largest one available" (Waelbrock, 1990: 20, cited in Ziltener, 2004: 956).

23 Gill noted that the phrase "locking in" was taken from the World Bank, *World Development Report: The State in a Changing World* (World Bank, 1997: 51), (Gill, 1998: 25).

24 For the root cause of the Euro sovereign debt crisis and the handling of the crisis by the SCAMD elites, see De Cecco et al., 2012; Stiglitz, 2015a, 2015b, 2016; Stiglitz et al. 2015; Skidelsky, 2015; Piketty et al., 2015; Krugman, 2015 a, 2015b, 2015c, 2015d, 2015e; Varoufakis, 2015a, 2015b, 2015c; Flassbeck and Lapavitsas, 2015; Flassbeck and Spiecker, 2011.

25 A system that allows polluters to avoid cutting emissions by paying money to allegedly improve someone else's emissions performance.

26 Clean Development Mechanism, a system sanctioned by the UN that turns claims of helping poor countries to reduce emissions into credits that can be traded on the market.

27 Emissions trading system (ETS). The EU's ETS is the world's first major carbon market and remains the largest one.

28 Anderson criticized that the process for determining the target was "conducted in a vacuum of scientific evidence" and relies on "the abuse of probabilities of 2°C" (Anderson, 2013).

29 Bergesen et al. noted that the choice of the goal was not random given that several of the member states had already set national emissions reduction targets. As a result, it appeared that stabilizing CO_2 emissions at 1990 levels by the year 2000 "would be feasible" (Bergesen et al., 1994: 17).

30 "TENs is the largest transport infrastructure plan in history. It includes a number of built and unbuilt monsters: the Channel Tunnel, the Øresund Bridge connecting Denmark and Sweden, a series of high-speed train links, numerous airport expansions and 12,000 kilometers of new motorways." The attraction of the TENs is obvious for oil companies such as BP, Petrofina, Shell, and Total, as well as car manufacturers such as DaimlerChrysler, Fiat, and Renault (Balanyá et al., 2003: 22–3, 69).

31 The author of the manifesto was the Business Council for Sustainable Development (BCSD), a body created by Swiss businessman Stephan Schmidheiny, who participated in Bilderberg between 1993 and 1995. Among BCSD's members were Volkswagen, Chevron, Shell, DuPont, and Dow (Cowe, 1992). It was the secretary general of the Rio Summit, Maurice Strong, an oilman, who asked Schmidheiny to "gather together a group of industrialists" which gave birth to the BCSD (Vidal, 1992). According to Vidal, the 370-page business manifesto entitled "Changing Course" was full of rhetoric, with "no sense that the great minds of world industry want anything to change except their profit margins."

32   John Dales and Thomas Crocker, economists who pioneered the concept of permit trading in the 1960s, suggested that states would be better placed than an imaginary perfect market to set a cap on overall pollution levels. Specifically regarding carbon emissions reduction, even these economists expressed skepticism about the effectiveness of using permit trading, since there were lots of situations in which their theory would not apply (Gilbertson and Reyes, 2009: 19).

33   Exxon Mobil Corporation 2007 Worldwide Contributions and Community Investments; SourceWatch: www.sourcewatch.org/index.php/Center_for_Clean_ Air_Policy#cite_note-2; Exxon Mobil Corporation 2014 Worldwide Contributions and Community Investments, cdn.exxonmobil.com/~/media/global/files/ worldwide-giving/2014-worldwide-contributions-public-policy.pdf.

34   Founded in 1989 and had financial support from the Ford Foundation (FIELD, 1993).

35   According to the *Oxford English Dictionary*, "shadow price" is "the estimated price of a good or service for which no market price exists."

36   Over the years environmentalists increasingly see the UN climate talks themselves as a contributing cause of climate crisis, with negotiators squabbling over calculations about emissions claim, rather than talking about ways to make deep cuts (The Corner House, 2001: 1).

37   A 2011 factsheet, for instance, was entitled: "The EU ETS is delivering emission cuts." Similarly, declaring that, for the 1990–2015 period, emissions decreased by 22% even when the EU's combined GDP grew by 50%, Climate Action and Energy Commissioner Miguel *Arias Cañete boasted triumphantly that* "once again we have shown that protecting the climate can go hand in hand with economic growth" (European Commission, 2016).

38   "The sectors at highest risk of relocating their production outside the EU will receive *full free allocation.* The free allocation rate for sectors less exposed to carbon leakage will amount to 30%" (Council of the EU, 2018; emphasis added).

39   The back-loading measure postponed the auctioning of 900 million allowances glutted between 2014 and 2016 until 2019–20 to rebalance supply and demand resulted from over-allocation (EU, 2014). The Market Stability Reserve was set up for the 900 million allowances back-loaded in 2014–2016 to be transferred to rather than auctioned in 2019–20 (EU 2015). The new measure devised to make the ETS work started operating in January 2019, one and a half decades after the launch of the scheme.

40   The EU Court of Auditors complains that, even after a €5 billion VAT fraud and scandals involving stolen and re-used allowances, the financial aspect of the ETS still remains under regulated (Corporate Europe Observatory, 2015).

References

Agnelli, Giovanni. (1971). The Contributions of Business in Dealing with Current Problems of Social Instability. Bilderberg Meeting. 23–25 April. Yale University Library Digital Repository. Henry A. Kissinger papers, part II.

Aldrich, Richard J. (1997). "OSS, CIA and European Unity: The American Committee on United Europe, 1948–1960," *Diplomacy & Statecraft*, Vol. 8, No. 1, 184–227.

Alter, Karen J. and Sophie Meunier-Aitsahalia. (1994). "Judicial politics in the European Community—European Integration and the pathbreaking Cassis de Dijon Decision," *Comparative Political Studies*. Vol. 26, no. 4, 535–561.

Anderson, Kevin. (2013). Open Letter to the EU Commission President About the Unscientific Framing of Its 2030 Decarbonization Target. *Kevinanderson.info*. 13 December.

Anderson, Kevin. (2017). A Précis of My Take on Our Collective "Stupidity." *Kevinanderson.info*. 24 July.

Anderson, Kevin. (2018). "Climate Change: Can the Environment Be Saved by Today's Emerging Political Forces?" *Independent*. 21 March.

Aubourg, Valerie. (2003). "Organizing Atlanticism: the Bilderberg Group and the Atlantic Institute, 1952–1963," *Intelligence and National Security*. 18(2): 92–105.

Balanyá, Belén, Ann Doherty, Olivier Hoedeman, Adam Ma'anit and Erik Wesselius. (2003). *Europe Inc. Regional & Global Restructuring & the Rise of Corporate Power*. London: Pluto Press.

Baumert, Kevin, Timothy Herzog and Jonathan Pershing. (2005). "Navigating the Numbers Greenhouse Gas Data and International Climate Policy." *World Resources* Institute.

BBC. (2005). Inside the Secretive Bilderberg Group. 29 September.

BBC. (2011). Bilderberg Mystery: Why Do People Believe in Cabals? 8 June.

Beddington-Behrens, Edward. (1960). "Mr. Joseph Retinger," *The Times*. 13 June, 12.

Bel, Germà and Stephan Joseph. (2015). "Emission Abatement: Untangling the Impacts of the EU ETS and the Economic Crisis, "*Energy Economics*. 49: 531–539.

Bergesen, Helge Ole, Michael Grubb, Jean-Charles Hourcade, Jill Jaeger, Alessandro Lanza, Reinhard Loske, Liv Astrid Sverdrup, and Angelica Tudini. (1994). *Implementing the European CO_2 Commitment—A Joint Policy Proposal*. London: The Royal Institute of International Affairs.

Bertell, Rosalie. (1996). "An Elite Pow Wow," *Peace Magazine*. Vol. 12, no. 4: 6.

Bird, Kai. (1992). John J. *The Chairman: McCloy—The Making of the American Establishment*. New York: Simon & Schuster.

Brown, Paul and Roger Cowe. (1997). "For Sale: The Right Not to Use This Factory," *The Guardian*. 11 December.

Buchanan, James and Gordon Tullock. (1962). *The Calculus of Consent: Logical Foundations of Constitutional Democracy*. Indianapolis: Liberty Fund.

Burbach, Roger. (2006). "Augusto and Us," *The Guardian*. 11 December.

Campbell, Alan. (2000). "Lord Greenhill of Harrow," *The Guardian*. 11 November.

Carbon Market Watch. (2016). "Industry Windfall Profits from Europe's Carbon Market 2008–2015," Carbon Market Watch Policy Briefing, November.

Carrol, William K. and Colin Carson. (2003). "Forging A New Hegemony? The Role of Transnational Policy Groups in the Network and Discourses of Global Corporate Governance." *Journal of World-Systems Research*. Vol. 9, no. 1, 67–102.

CCAP. (1999). Design of A Practical Approach to Greenhouse Gas Emissions Trading Combined with Policies and Measures in the EC. Washington, DC: Center for Clean Air Policy.

CE Delft. (2016). Calculation of Additional Profits of Sectors and Firms from the EU ETS.

Comello, Stephen and Stefan Reichelstein. (2014). A Case Study in Internal Carbon Pricing: Royal Dutch Shell. Corporate Use of Carbon Prices. White Paper, Carbon Disclosure Project.

Commission of European Communities. (1985a). Program of the Commission for 1985. Bulletin of the European Communities Supplement 4/85. http://aei.pitt.edu/7630/1/31735055261907_1.pdf.

Commission of the European Communities. (1985b). Completing the Internal Market. White Paper from the Commission to the European Council (Milan, 28–29 June). COM(85) 310 final. Brussels, 14 June.

Cooper, Richard. (1994). Yes to European Monetary Unification, but No to the Maastricht Treaty. In *Alfred Steinherr* (ed.). *Thirty Years of European Monetary Integration*. London: Longman.

Corporate Europe Observatory. (2015). EU Emissions Trading: 5 Reasons to Scrap The ETS. 26 October.

Council of the EU. (2018). EU Emissions Trading System Reform: Council Approves New Rules for the Period 2021 to 2030. Press Release. 27 February.

Cowe, Roger. (1992). "Business Wakes Up to the Environment," *The Guardian*. May 8.

Cowles, Maria Green. (1995). "Setting the Agenda for a New Europe: The ERT and EC 1992," *Journal of Common Market Studies*. Vol. 33, no. 4, 501–526.

Davies, Nick. (2007). "The Inconvenient Truth about the Carbon Offset Industry," *The Guardian*. June 16.

Davignon, Etienne. (1982). Answer given by Mr. Davignon on behalf of the Commission. 30 November. *Official Journal of the European Communities*. No C3/15. 5 January 1983.

De Cecco, Marcello et al. (2012). "Fixing Trade Imbalances Is Only Way to Avoid Eurozone Implosion," *Financial Times*. 23 January.

Dekker, Wisse. (1985). "Europe 1990—An Agenda for Action," *European Management Journal*. Vol. 3, no. 1, 5–10.

Di Meana, Carlo Ripa. (1992). "Why I'm No in Rio," *The Observer*. 7 June.

The Economist. (1969). "The Giants' Causeway." 27 December: 10.

The Economist. (1992a). "The Green Man in Brussels." 25 April.

The Economist. (1988). "Europe's Internal Market—After the fireworks." 9 July: 5–12.

The Economist. (1992b). "Europe's Industries Play Dirty." 9 May.

The Economist. (2000). "Economics Focus: The Ethics Gap." 2 December.

The Economist. (2008). "Billion Dollar Babies." 26 April.

Engdahl, William. (2004). *Century of War: Anglo-American Oil Politics and the New World Order*. London: Pluto Press.

ERT. (1994). The Climate Change Debate—Seven Principles for Practical Policies. December.

ERT. (1997). Climate Change—An ERT Report on Positive Action. December.

EURACTIV. (2017). "The EU: What Happened to Climate's Poster Child?" 17 May.

European Commission. (1985). "Proposal for a Council Decision adopting three multiannual research and development programs in the field of the environment (1986 to 1990) (Environmental protection, Climatology, Major technological hazards)." *Official Journal of the European Communities*. C 301. Vol. 28. 25 November.

European Commission. (1988). Communication to the Council on The Greenhouse Effect and the Community: Commission Work Programme Concerning the Evaluation of Policy Options to Deal with the "Greenhouse Effect." COM 88(656) final. 16 November.

European Commission. (1991). Communication to the Council on A Community Strategy to Limit Carbon Dioxide Emissions and to Improve Energy Efficiency. SEC(91)1744 final. 14 October.

European Commission. (1992). Proposal for a Council Directive Introducing a Tax on Carbon Dioxide Emissions and Energy. COM(92)226 final. 30 June.

European Commission. (1998). Communication on Climate Change—Towards An EU Post-Kyoto Strategy. COM(1998)353 final. 3 June.

European Commission. (2000). Green Paper on Greenhouse Gas Emissions Trading within the European Union. COM(2000)87 final. 8 March.

European Commission. (2016). Report on Implementing the Paris Agreement—Progress of the EU towards the at Least -40% Target. COM(2016) 707 final.

European Commission. (2017a). Report on the Functioning of the European Carbon Market. COM(2017)693 final. 23 November.

European Commission. (2017b). Report on the Functioning of the European Carbon Market. COM(2017)48 final. 1 February.

European Council. (1981). Meeting of the European Council in Maastricht, *European Community News.* No. 11/1981. 25 March.

European Council. (1990). Presidency Conclusions. Dublin, 25 and 26 June 1990. SN 60/1/90.

European Council. (2014). European Council Conclusions. (23 and 24 October 2014) EUCO 169/14.

European Parliament. (1986). Resolution on Measures to Counteract the Rising Concentration of Carbon Dioxide in the Atmosphere (The 'Greenhouse' Effect). *Official Journal of the European Communities.* C 255. 13 October, 272.

European Union. (2014). Commission Regulation (EU) No 176/2014 of 25 February amending Regulation (EU) No 1031/2010 in particular to determine the volumes of greenhouse gas emission allowances to be auctioned in 2013–2020.

European Union. (2015). Decision (EU) 2015/1814 of the European Parliament and of the Council of 6 October 2015 concerning the establishment and operation of a market stability reserve for the Union greenhouse gas emission trading scheme and amending Directive 2003/87/EC.

Facchinetti, Ronald. (1994). "Why Brussels Has 10,000 Lobbyists," *The New York Times*, 21 August.

FIELD. (1993). Sustainable Development: The Challenge to International Law—Report of a Consultation Held at Windsor 27 to 29 April 1993. *Review of European, Comparative & International Environmental Law.* Vol.2(4): r1-r15.

FIELD. (2000). Designing Options for Implementing and Emissions Trading Regime for Greenhouse Gases in the EC. 22 February.

Financial Times. (1991). "The Greening of Taxation," 2 October.

Flassbeck, Heiner and Costas Lapavitsas. (2015). *Against the Troika: Crisis and Austerity in the Eurozone.* London: Verso Books.

Flassbeck, Heiner and Friederike Spiecker. (2011). "The Euro—a Story of Misunderstanding," *Intereconomics* 4: 180–187.

Gardner, David. (1991a). "EC Ministers Welcome Energy Tax Idea." *Financial Times.* 2 October.

Gardner, David. (1991b). "EC Energy Tax Moves A Stage Closer." *Financial Times.* 14 December.

Gardner, James N. (1991a). "Lobbying, European-Style." *Europe.* November, 29–30.

Gardner, James N. (1991b). *Effective Lobbying in the European Community.* Boston: Kluwer Law and Taxation Publishers.

Giersch, Herbert. (1985). Eurosclerosis. Lecture delivered in Sydney on 20 August 1985 at the Regional Meeting of the Mont Pelerin Society.

Gilbertson, Tamra and Oscar Reyes. (2009). Carbon Trading—How it Works and Why It Fails. Uppsala: Dag Hammarskjöld Foundation.

Gill, Stephen. (1990). *American Hegemony and The Trilateral Commission*. Cambridge: Cambridge University Press.

Gill, Stephen. (1998). "European Governance and New Constitutionalism: Economic and Monetary Union and Alternatives to Disciplinary Neoliberalism in Europe," *New Political Economy*. 3(1): 5–26.

Gormley, Laurence. (1981). "Cassis de Dijon and the Communication from the Commission," *European Law Review*. Vol. 6, 454–459.

Grahl, John and Paul Teague. (1989). "The Cost of Neoliberal Europe," *New Left Review*. 1/174, March–April, 33–50.

Groenleer, Martijn L.P. and Louise G. van Schaik. (2007). "United We Stand? The European Union's International Actorness in the Cases of the International Criminal Court and the Kyoto Protocol." *Journal of Common Market Studies*, 45(5): 969–998.

Gupta, Joyeeta and Michael Grubb (eds.). (2000). *Climate Change and European Leadership*. Dordrecht: Kluwer Academic Publishers.

Hanson, Brian T. (1998). "What Happened to Fortress Europe? External Trade Policy Liberalization in the European Union," *International Organization*. 52(1), 55–85.

Hatch, Alden. (1962). *Prince Bernhard of the Netherlands*. London: George G. Harrap & Co.

Hu, Jing, Wina Crijns-Graus, Long Lam and Alyssa Gilbert. (2015). "Ex-ante Evaluation of EU ETS during 2013–2030: EU-Internal Abatement," *Energy Policy*. 77: 152–163.

Hunt, John. (1991a). "EC Energy Tax Idea Could Add $10 to a Barrel of Oil," *Financial Times*. 5 February.

Hunt, John. (1991b). "Industrialists Challenge Energy Tax," *Financial Times*. 24 September.

Hutton, Will. (1992). "Heated Debate on Road to Rio," *The Guardian*. 18 May.

Jack, Andrew. (1991). "Scrivener Caveat on Energy Tax Idea," *Financial Times*. 8 November.

Klein, Naomi. (2014). *This Changes Everything—Capitalism vs. The Climate*. New York: Simon & Schuster.

Krugman, Paul. (2015a). "Ending Greece's Nightmare," *The New York Times*. 26 January.

Krugman, Paul. (2015b). "What Greece Won," *The New York Times*. 27 February.

Krugman, Paul. (2015c). "Notes on Greece," *The New York Times*. 19 April.

Krugman, Paul. (2015d). "Greece on the Brink," *The New York Times*. 20 April.

Krugman, Paul. (2015e). "Ending Greece's Bleeding," *The New York Times*. 5 July.

Laing, Timothy, Misato Sato, Michael Grubb, & Claudia Comberti. (2014). "The Effects and Side-Effects of the EU Emissions Trading Scheme," *Climate Change*. Vol. 5, issue 4.

Lean, Geoffrey and Polly Ghazi. (1992). "Last-Minute Deal Dilutes Pledge on Global Warming," *The Observer*. 10 May.

Lewis, Paul. (1995). "Emilio Collado, a Creator of the World Money System, Dies at 84," *The New York Times*. 16 February.

Ma'anit, Adam. (2000). Greenhouse Market Mania—UN Climate Talks Corrupted by Corporate Pseudo-solutions. Report published by Corporate Europe Observatory. A summarized and updated version available at www.twn.my/title/twr125e.htm.

MacLean, Nancy. (2017). *Democracy in Chains: The Deep History of the Radical Right's Stealth Plan for America*. New York: Penguin Publishing Group.

Marshall, Andrew Gavin. (2009). "Controlling the Global Economy: Bilderberg, the Trilateral Commission and the Federal Reserve," *Global Research*. 3 August.

Martin, Douglas. (2007). "Robert O. Anderson, Oil Executive, Dies at 90," *The New York Times*. 6 December.

Moorehead, Caroline. (1977). "An exclusive club, perhaps without power, but certainly with influence: The Bilderberg group," *The Times*. 18 April: 9.

Morris, Damien. (2013). *Drifting toward Disaster? The ETS Adrift in Europe's Climate Efforts*. London: Sandbag.

Nollert, Michael and Nicola Fielder. (2000). "Lobbying for A Europe of Big Business: The European Roundtable of Industrialists." In Volker Bornschier (ed.), *State-Building in Europe—The Revitalization of Western European Integration*. Cambridge: Cambridge University Press, 187–209.

Oberthür, Sebastian and Claire Roche Kelly. (2008). "EU Leadership in International Climate Policy: Achievements and Challenges," *International Spectator*. 43(3): 35–50.

Oberthür, Sebastian and Marc Pallemaerts. (2010). "The EU's Internal and External Climate Policies: An Historical Overview." In Oberthür, Sebastian and Marc Pallemaerts (eds.), *The New Climate Policies of the European Union—Internal Legislation and Climate Diplomacy*. Brussels: Brussels University Press, 27–64.

Open Letter from European Economists to the Heads of Government of the 15 Member States of the European Union. (1997). www.hartford-hwp.com/archives/60/068.html.

Padoa-Schioppa, Tommaso. (1987). *Efficiency, Stability and Equity: A Strategy for the Evaluation of the Economic System of the EC*. Oxford: Oxford University Press.

Palmer, John. (1992a). "EC Wavers over Plan to Tax Carbon Dioxide Emissions," *The Guardian*. 30 April.

Palmer, John. (1992b). "EC Plans Carbon Fuel Tax," *The Guardian*. 14 May.

Peters, Glen P., Jan C. Minx, Christopher L. Weber and Ottmar Edenhofer. (2011). "Growth in Emission Transfers via International Trade from 1990 to 2008," *Proceedings of the National Academy of Sciences of the United States of America*. Vol. 108, No. 21: 8903–8908.

Pieczewski, Andrzej. (2010). "Joseph Retinger's conception of and contribution to the early process of European integration," *European Review of History*. 17(4): 581–604.

Piketty, Thomas et al. (2015). "Austerity Has Failed: An Open Letter from Thomas Piketty to Angela Merkel," *The Nation*. 7 July.

Retinger, Joseph. (1956). *The Bilderberg Group*. https://publicintelligence.net/bilderberg-group-retinger/.

Rettman, Andrew. (2009). " 'Jury's Out' on Future of Europe, EU Doyen Says," *EUobserver*. 16 March.

Reyes, Oscar and Belén Balanyá. (2016). Carbon Welfare—How Big Polluters Plan to Profit from EU Emissions Trading Reform. Corporate Europe Observatory Report.

Reyes, Oscar. (2014). *Life Beyond Emissions Trading*. Corporate Europe Observatory.

Rocha, Marcia, Mario Krapp, Johannes Guetschow, Louise Jeffery, Bill Hare and Michiel Schaeffer. (2015). Historical Responsibility for Climate Change–from Countries Emissions to Contribution to Temperature Increase. Climate Analytics Report.

Sandholtz, Wayne and John Zysman. (1989). "1992: Recasting The European Bargain," *World Politics*. 42(1): 95–128.

Schmidt, Susanne K. (2007). "Mutual Recognition As A New Mode of Governance," *Journal of European Public Policy*. 14(5): 667–681.

Schreurs, Miranda A. and Yves Tiberghien. (2007). "Multi-Level Reinforcement: Explaining European Union Leadership in Climate Change Mitigation," *Global Environmental Politics*. 7(4): 19–46.

Shell. (2002). Meeting the Energy Challenge—The Shell Report 2002.

Skelton, Charlie. (2018). "Bilderberg 2018: New Tech Helps Oil the Wheels of the Global Elite," *The Guardian*. 7 June.

Skidelsky, Robert. (2015). "I Agree with Syriza: The Way Back to Prosperity Is Not Austerity but Debt Relief," *New Statesman*. 6 February.

Skjaersethrseth, Jon Birger. (1994). "The Climate Policy of the EC: Too Hot to Handle?" *Journal of Common Market Studies*. 32(1), 25–45.

Stiglitz, Joseph et al. (2015). "EU Leaders' Chance to Create Good History," *Financial Times*. 29 June.

Stiglitz, Joseph. (2015a). "Europe's Lapse of Reason," *Project Syndicate*. 8 January.

Stiglitz, Joseph. (2015b). "How I Would Vote in the Greek Referendum," *The Guardian*. 29 June.

Stiglitz, Joseph. (2016). *The Euro: How A Common Currency Threatens the Future of Europe*. New York: W. W. Norton & Company.

Talani, Leila Simona. (2003). The Political Economy of Exchange Rate Commitments—Italy, the United Kingdom, and the Process of European Monetary Integration. In Alan W. Cafruny and Magnus Ryner (eds.), *A Ruined Fortress? Neoliberal Hegemony and Transformation in Europe*. Lanham, MD: Rowman & Littlefield Publishers, 123–146.

The Corner House. (2001). Democracy or Carbocracy? Intellectual Corruption and the Future of the Climate Debate.

Thompson, Peter. (1980). Bilderberg and the West. In Holly Sklar (ed.) *Trilateralism—The Trilateral Commission and Elite Planning for World Management*. Boston: South End Press. 157–189.

Time. (1972). "Executives: Wagnerian Era," 17 July.

UNICE. (1991). Position on a Number of Basic Principles for the Formulation of a Community Action Strategy on the Greenhouse Effect. 25 June.

Van Apeldoorn, Bastiaan. (2002). *Transnational Capitalism and the Struggle over European Integration*. New York: Routledge.

Van Apeldoorn, Bastiaan. (2019). Transnational Class Agency And European Governance. In Bob Jessop and Henk Overbeek (eds.), *Transnational Capital And Class Fraction*. London: Routledge, 119–144.

Van Der Pijl, Kees. (2012). *The Making of An Atlantic Ruling Class*. London: Verso Books.

Varoufakis, Yanis. (2015a). Greece's Proposals to End the Crisis: My Intervention at Today's Eurogroup. 18 June. Yanis Varoufakis's personal blog: thoughts for the post-2008 world. http://yanisvaroufakis.eu/2015/06/18/greeces-proposals-to-end-the-crisis-my-intervention-at-todays-eurogroup/. Latest update 18 June 2015.

Varoufakis, Yanis. (2015b). Speech at the Panel Session "Greece's Future in the EU" at the Hans-Boeckler foundation in Berlin, 8 June.

Varoufakis, Yanis. (2015c). "Greece and Dr. Schäuble's Plan for Europe: Do Europeans Approve?" *Global Research*. 22 July.

Vidal, John. (1992). "A World Shackled by Economic Chains," *The Guardian*. 8 May.

Waelbrock, Jean. (1990). 1992: Are the Figures Right? Reflections of a Thirty Per Cent Policy Maker. In H. Siebert (ed.) *The Completion of the Internal Market*. Tübingen: Mohr Siebeck.

Watkins, Kevin. (1992). "The Foxes Take Over the Hen House," *The Guardian*. 17 July.

Welch, Dan. "A Buyer's Guide to Offsets," *Ethical Consumer*. 106.

Wilford, Hugh. (2003). "CIA Plot, Socialist Conspiracy, or New World Order? The Origin of the Bilderberg Group, 1952–1955," *Diplomacy and Statecraft*. 14(3): 70–82.

Wolf, Julie. (1991). "Support Within EC for an Energy Tax to Combat Global Warming Is Growing," *The Wall Street Journal*. 16 December.

Wolf, Julie. (1992). "EC Divided over Stance for Rio," *The Guardian*. 27 May.

World Bank. (1997). World Development Report 1997: The State in a Changing World.

World Bank. (2017). Report of the High-Level Commission on Carbon Prices.

Ziltener, Patrick. (2004). "The Effects of European Integration on Economic Growth and Convergence of its Member Countries," *Review of International Political Economy*. Vol. 11, no. 5, 953–979.

6 The SCAMD's Judicial Branch—
The Investor-State Dispute Settlement

The success of the SCAMD elites in sabotaging carbon/energy tax proposals and installing self-serving carbon trading systems leaves government regulation the only remaining solution for effective climate mitigation. Despite incessant neoliberal efforts to deregulate, new regulations continue to be issued in response to new scientific findings on environmental, health, and other public interest-related issues. To counter these developments, SCAMD elites resorted to building an international arbitration regime (Bonnitcha, 2017: 2) catering exclusively to the needs of multinationals, with the power to override national executive, legislative, and judicial decisions. Governments daring to protect the public interest in ways that undercut the interests of foreign investors can be brought to neoliberal justice through a system known as Investor-State Dispute Settlement (ISDS). As Stone Sweet and Grisel noted, "by the end of the twentieth century... the tendency to keep transnational commercial disputes out of the courts, and thereby beyond the reach of local laws, is nearly universal" (2017: 1). Like the EU's emissions trading scheme (ETS), the ISDS fends off mitigation measures while awarding money to big fossil fuel companies for polluting. Given how well the system has worked, its continual expansion remains high on the SCAMD agenda. Such an expansion through treaties such as the Energy Charter Treaty (ECT), the EU-Canada Comprehensive Economic and Trade Agreement (CETA), the Transatlantic Trade and Investment Partnership between the EU and the U.S. (TTIP), the Comprehensive and Progressive Agreement for Trans-Pacific Partnership (CPTPP),[1] and other multilateral or bilateral investment agreements will be exceedingly effective for locking climate action behind an impassable moat. The first multilateral treaty through which the big fossil fuel companies had consolidated ISDS and disseminated its application is the ECT. This chapter begins with the history of the ECT and its on-going damage before moving to discuss the overall neoliberal assault on climate regulations through further expansion and entrenchment of the ISDS regime.

In May 2017 the British-based oil and gas company Rockhopper sued the Italian government in an investment tribunal for a ban on offshore oil drilling near the Italian coast. The ban, which came only after a decade-long protest by local residents, was issued on the grounds of environmental concerns and high earthquake risks. The compensation that Rockhopper demanded covered not just what it has actually spent for exploring an oilfield in the Adriatic Sea but also an additional \$200–300 million in hypothetical profits that the company believed could have been generated, had the Italian government not issued the ban (Eberhardt et al., 2018: 14). In 2009 Swedish energy multinational Vattenfall sued Germany for €1.4 billion in compensation for environmental restrictions imposed on a coal-fired power plant near Hamburg. Vattenfall argued that the new standards set by local authorities would render the company's coal-fired power plant "uneconomical" (Vattenfall, 2009: 12). Faced with the intimidating €1.4 billion claim in compensation, the local authorities settled with Vattenfall with a smaller amount of compensation, while also relaxing the environmental restrictions (Eberhardt et al., 2018: 22–3). In 2012 Vattenfall filed a second investment arbitration against Germany for \$5.4 billion, for German Chancellor Angela Merkel's decision to phase out nuclear power as a result of the nuclear disaster at Fukushima in Japan in 2011 (Deutsche Welle, 2019).

All three cases were filed according to the same treaty, i.e., the ECT. The Energy Charter political declaration was signed in The Hague in February 1992—the same month that the Treaty of Maastricht was signed. The founding father of the ECT, Ruud Lubbers, was also a key figure behind the Maastricht Treaty and a frequent Bilderberg participant in the critical period when both Treaties were being negotiated and signed. A former businessman active in employers' associations, Lubbers became the longest-serving prime minister of the Netherlands and was often compared to Margaret Thatcher and Reagan Reagan, given his familiar slogan, "more markets, less government" (Energy Charter Secretariat, 2018; Van der Vat, 2018). At the June 1990 EC summit—where climate change was put on the agenda for the first time—Lubbers proposed a European Energy Network to protect European energy investment abroad (European Council, 1990: 13). "His vision and political drive soon led to the founding 'bricks and mortar' of the Energy Charter Process, eventually leading to nearly 50 countries signing the Energy Charter Treaty in December 1994" (Energy Charter Secretariat, 2018). According to the European Commission press release for the signing of the ECT in December 1994, the EC had from the very start played a major role in the Treaty not only as a coordinator, but also as "driving force behind the negotiations" that accelerated the process when the process slowed down." In fact, the Treaty itself was "drafted by the European Commission in November 1990," with the Commission also hosting the Secretariat in one of its buildings (European Commission, 1994).[2]

The aim of the ECT, as the Commission explained in its press release, was to provide greater legal certainty for energy-related investments. "The Treaty recognizes State sovereignty over energy resources, *but* Contracting Parties commit themselves to facilitate access to resources as well as transit of energy" (European Commission, 1994; emphasis added). As a result, the Treaty envisaged arbitrage procedure for settling investment disputes. During the negotiations, large oil and gas companies, including Shell, Exxon Mobil, ENI, BP, and Repsol remained closely involved. As well as being invited to comment on draft versions of the future treaty, fossil fuel lobby groups such as the Exploration and Production Forum,[3] Eurogas, and Europia[4] also had regular meetings with lead negotiators. Their instruction for the negotiators was essentially that the Treaty must guarantee the protection of investments and give "investors the right to enforce certain key provisions directly against the host government through arbitration" (Eberhardt et al., 2018: 15). Given the origin of the Treaty, it is unsurprising that, as of November 2017, out of the 42 companies sitting on the Industry Advisory Panel of the ECT, at least 36 make money from fossil fuels and have had appalling track records for driving climate change (Eberhardt et al., 2018: 39). The EU, the ECT Secretariat, and the arbitration industry continue to aggressively promote the expansion of the ECT, cajoling countries endowed with rich energy resources to join the Treaty. The line of argument used to encourage such countries, which tend to lack awareness about the risks, to sign up includes "clean energy investment for all" and the Treaty as the pre-condition for attracting foreign investment (Eberhardt et al., 2018: 4).[5]

The international tribunals entrusted with the enormous power of ordering defendant states to pay billions of dollars of taxpayer money to compensate foreign investors for their claimed loss consist of only three for-profit private corporate lawyers. These arbitrators are often biased in favor of investors who could, outside the case at hand, also be their clients. The interchangeable role between arbitrator and corporate lawyer in different cases explains why merely 25 arbitrators have captured 44% of the ECT ISDS cases, and two-thirds of them have acted as legal counsel in other investment disputes. In addition, the fact that ISDS arbitrators are handsomely paid (roughly $3,000 a day) creates a financial incentive to rule in favor of investors, who are the only parties eligible[6] to bring claims to the tribunal (Eberhardt et al., 2018: 3, 18). In a bundle of related ISDS cases commonly referred to as *Yukos v. Russia* filed under the ECT, the arbitrators ordered Russia to pay $50 billion compensation and 83% of the $124 million legal costs. Apart from the $1,065 per hour charged by Yukos's lawyers, the legal costs covered the €5.3 million that went to the tribunal's three arbitrators and nearly €1 million that went to their assistant (Eberhardt et al., 2018: 25–6). In fact, ISDS arbitrations are so lucrative that third-party funders who finance the legal costs in exchange for a share in any granted award or settlement are becoming increasingly common (Eberhardt et al., 2018: 3). In *Rockhopper v. Italy*, for instance, the company announced it had "secured effectively no-win, no-fee funding for the case from a specialist in financing commercial litigation and arbitration claims" (Gosden, 2017). These third-party funders are often speculators who "can profit if the government is ordered to pay massive damages, but can't be required to pay the costs if the case is lost." As a result, the government has to pay when it loses but may not get paid when it wins and the foreign investor is ordered to pay. The phenomena of third-party funders combined with the fact the foreign investors can be shell companies with no assets leads to the shocking statistics that governments fail to recover their legal costs in 37% of cases (PSI, 2018).

This predatory SCAMD judicial system leaves climate action extremely vulnerable. By definition, climate action entails installing *new* regulations that place the planet before private corporate interests. Such re-prioritization is exactly what the ISDS is being used to protect investors from. Sunset industries that are harmful to public interests are becoming a goldmine for ISDS speculators. ISDS arbitrators have typically invoked "fair and equitable treatment" (FET) for their decisions to award billions of dollars to investors, even though the concept, with its traditional meaning of the minimum standard of treatment for foreign nationals in customary international law, is generally deferential to a country's right to regulate (Van Harten, 2016: 3).[7] ISDS arbitrators have stretched the meaning of FET so much that it could imply that governments have an obligation to *make no changes* in so far as such changes could undercut interests for foreign investors. In *Eiser v. Spain*, a case again based on the ECT, the ISDS arbitrators awarded €128 million to the investor in May 2017, arguing that by radically altering regulations, the Spanish government "crossed the line" and "violated the obligation to accord fair and equitable treatment," frustrating the expectations of the investor (Ross, 2017; Eberhardt et al., 2018: 18, 46).

While the ECT is the treaty that has triggered the most ISDS lawsuits in the world—114 as of June 2018[8]—and has been the most heavily relied on by fossil fuel corporations, it is far from the only route through which fossil fuel companies have launched their ISDS attacks. The case of *Chevron & Texaco v. Ecuador* offers a glimpse of how Big Oil has used bilateral investment agreements to protect their climate-wrecking interests through ISDS. In August 2018 a three-person ISDS tribunal ruled that, owing to the signing of the US-Ecuador Bilateral Investment Treaty that went into effect in 1997 and contained ISDS clauses, by allowing its judicial system to rule against Chevron and Texaco for polluting its land, Ecuador committed international wrongs and denied justice to Chevron and Texaco.[9] The tribunal therefore ruled that Ecuador was liable to make reparations to Chevron and Texaco for injuries caused by the breaches of the fair and equitable treatment of standard and customary international law (AFTINET, 2018; Telesur, 2018).[10] Ecuador is a victim of environmental pollution resulting from the dumping of 16 billion gallons of toxic waste into waterways between 1964 and 1992 by Texaco, acquired in 2001 by Chevron, affecting more than 30,000 indigenous people (Business & Human Rights Resource Centre, 2018). The land now known as "Amazonian Chernobyl" was once pristine rainforest (Kendall, 2008). Among the six indigenous groups—the Cofán, Secoya, Sionas,

Huaorani, Quichua, and Tetete—affected, the Tetete reportedly became extinct shortly after Texaco started operating in the area, and the Cofán in the area lost 74% of its population (CETIM, 2015; Chevron in Ecuador, 2010). Despite Ecuador's role as a victim of reckless pollution, it already made a payment of $112 million to Chevron in 2016, executing the ISDS tribunal's Interim Measures Order issued in 2011 ordering Ecuador to suspend enforcement of its domestic judgment and awarding Chevron $96 million (the $112 million consists of the $96 million award plus interest) (Business & Human Rights Resource Centre, 2018).

From the huge pile of other similar cases,[11] the ISDS claim brought by TransCanada against the Obama Administration's rejection of the Keystone XL pipeline may have the most devastating climate implications. In May 2012 TransCanada applied to build an 875-mile pipeline that would deliver up to 830,000 barrels per day of crude oil extracted from tar sands in Alberta, Canada to the U.S. Gulf coast area. Not only would the pipeline cross more than a thousand U.S. rivers, streams, lakes, and wetlands, but the tar sands oil transported by the pipeline is also more corrosive than other forms of crude oil (Sierra Club, 2013: 1).[12] In order to be cost-effective, once built, the pipeline would force the dirty Canadian tar sands oil to be extracted and regularly exported to compete in the global markets (NRDC, 2015; Oil Change International, 2013). James Hansen simply explained the devastating effect of building the pipeline: "essentially, it's game over for the planet" (Mayer, 2011). In November 2015 the U.S. government denied TransCanada's application for the construction of the pipeline. Two months later, TransCanada filed notice of intent to start an ISDS against the U.S. government under the North American Free Trade Agreement (NAFTA). The company argued that the U.S. government had led it to develop "reasonable expectations" that the U.S. government would approve the pipeline, only to reject it in the end (TransCanada, 2016a). Treating the denial of the right to build the pipeline as a form of expropriation, TransCanada demanded—counting in the expected profits it estimated that it would have earned, had the project been approved—$15 billion in compensation when it formally filed the ISDS case in June 2016 (TransCanada, 2016b). After President Donald Trump signed an executive order to invite the company back to reapply for permits, among other concessions, the ISDS case was dropped.

During the renegotiation of NAFTA—the Treaty that entitled Canadian tar sands oil exporters to bring ISDS claims against the U.S.—in 2018, more than 300 state legislators from all 50 U.S. states and more than 1,000 civil society groups asked the U.S. trade negotiators to remove ISDS from the treaty. The American Legislative Exchange Council, which is funded by billionaires Charles and David Koch, asked to keep ISDS in the treaty (Citizens Trade Campaign, 2018). Unsurprisingly, it was the latter that got the ear of the Trump administration. The new treaty, the U.S.-Mexico-Canada Agreement (USMCA, the successor to NAFTA), which makes no mention of climate change, has made multinational energy companies the winners. While the treaty slowly phases out the ISDS between U.S. and Canada, it keeps the ISDS intact for the oil, gas, and power-generation sectors in the Mexican market, in which BP, Chevron, ExxonMobil, Shell, and Total have all invested (Grandoni, 2018). The USMCA fossil-fuels exemption from the elimination of ISDS is remarkable, given that the exact opposite action—the carving-out of all climate-wrecking industries from ISDS protections worldwide—is the only sensible thing to do short of a global, encompassing elimination of the ISDS. The ISDS carve-out already has a precedent. The CPTPP, a trade and investment agreement signed in 2018, linking 11 Asia-Pacific economies, contains an article preventing tobacco companies from bringing ISDS claims against government regulations on tobacco.[13] While having a tobacco carve-out in investment treaties is better than its absence, based on the evidence that climate-wrecking industries are by far the heaviest users of the ISDS—24% of all the cases registered with the ISCID up to 2018 concerned oil, gas, and mining, and another 17% concerned electric power and other energy resources—an ISDS carve-out for climate-related policies is even more urgent than tobacco carve-out (Tienhaara, 2017: 249; ISCID, 2019: 12; Van Harten, 2015).[14]

As a mechanism to channel money from taxpayers to large transnational corporations, the ISDS places no cap on the amount of awards governments can be ordered to pay. A review of 86 ISDS awards shows that countries have been ordered to transfer $9.2 billion of taxpayer money to foreign investors, of which $7.5 billion went to large[15] or extra-large[16] companies (Van Harten, 2016: 3). Like the EU ETS, the system turns "polluter pays" into "polluter gains." The gain, however, does not have to come in the form of compensation ordered by the tribunals. The chilling effect of the ISDS, which deters governments from adopting necessary regulations, brings in money by guaranteeing companies' right to pollute. According to the minutes of a 2014 meeting between Chevron executives and European Commission officials, Chevron argued that "the mere existence of ISDS is important as it acts as a deterrent" (Nelsen, 2016). Similarly, in a letter to the US Trade Representative in 2013, Chevron argued that "the existence of ISDS 'increases the likelihood' of disputes being settled outside them" (Tienhaara, 2017: 241). In the same vein, a Washington, D.C.-based law firm advises its clients in the energy sector to take advantage of the ISDS: "It may well be possible to use such protections as *a tool to assist lobbying* efforts to *prevent wrongful regulatory change*, or they may prove essential in obtaining compensation" (Coleman et al., 2014, emphasis added). As a result of the regulatory chilling effect, policymakers have to weigh the costs of being sued before policy proposals protecting public interests can even begin to take shape. The EU as a whole is already a target of ISDS deterrence. Investment lawyers have argued that the EU's renewable energy policies "have the potential to harm energy investments and may therefore violate the standards of protection promised to foreign investors in the ECT" (Eberhardt et al., 2018: 25). Owing to the fact that

ISDS rules are vaguely drafted and follow no system of precedent, the chilling effect is "much broader than other areas of law," leading governments to "shift away from a particular policy if there's any risk of a claim," as an experienced arbitration lawyer pointed out (Landau, 2014). While the Director General of the World Health Organization (WHO), Margaret Chan, has viewed the ISDS as "deliberately designed to instill fear in countries wishing to introduce... tough tobacco control measures" (Chan, 2012), the most catastrophic chilling effect that the ISDS has created no doubt lies with climate mitigation measures. This explains why the European Parliament has adopted a resolution requesting that "any measure adopted by a Party to the Paris Agreement...will not be subject to any existing or future treaty of a Party to the extent that it allows for investor-state dispute settlement" (European Parliament, 2015).

In defending the ISDS, its proponents often argue that the system is a necessary means for protecting investor interests, as it imposes "the rule of law in non-democratic states with a weak law and order tradition" (Schultz and Dupont, 2015: 1147). This fig leaf for an arrangement with neocolonial origins—an instrument for strengthening the economic interests of rich states at the expense of poor states—ran into problems when, starting in the mid-to-late 1990s, investors began to use the ISDS to attack rich and democratic states as well (Schultz and Dupont, 2015). The CETA signed by the EU and Canada in 2016 and the half-baked TTIP between the EU and the U.S. present the biggest problem for the line of argument that depicts the ISDS as a device for protecting investors from disrespect for democracy and the rule of law in backward countries. Testimonies by representatives of large corporations insisting on the inclusion of the ISDS in the EU-U.S. TTIP in the U.S. Senate reveal the treacherous nature of ISDS in *undermining* rather than *protecting* democracy and the rule of law. In a hearing on TTIP held in October 2013, Senator Sherrod Brown stated that "[t]hese [ISDS] provisions... do shift power, in reality, to a corporation to challenge a sovereign government." He then asked one of the witnesses:

> Does it concern you, as an American citizen, living in a country of laws and democratically reached rules, regulations, and statutes, to allow a foreign investor to, in essence, challenge, to have the standing to challenge, a democratically attained rule of law in this country?

The reply of the witness, Dave Ricks, senior vice-president of Eli Lilly, was "[i]t does not concern me, *as long as it is quid pro quo; as long as we have the same rights in their system.* And increasingly, companies like mine are global companies. We have an interest in many geographies" (Committee on Finance, U.S. Senate, 2013: 19 emphasis added). In other words, the U.S. should be relaxed about giving European corporations the right to attack democratically legislated laws in the U.S., as the arrangement is reciprocal: U.S. corporations will be empowered to attack democratically attained rules in Europe as well. Moreover, attacks going in both directions can actually come from the same corporations, given that ISDS protection is exclusively for transnationals, which by definition operate across countries.[17]

Like all the other neoliberal edifices, the TTIP was built top-down within the SCAMD structure. In 1995 the U.S. Department of Commerce and the European Commission convened the Trans-Atlantic Business Dialogue (TABD) to "serve as the official dialogue between American and European business leaders and U.S. cabinet secretaries and EU commissioners." According to its website,[18] the TABD brings together "chief executive officers and C-Suite executives from leading American and European companies operating in the U.S., Europe, and globally who advocate for a barrier-free transatlantic market." To achieve this goal,

> TABD provides its member executives *high-level access to U.S. Cabinet Secretaries and European Commissioners*. It has traditionally provided a set of policy recommendations important to business ahead of each U.S.-EU Summit; and on occasion, presented these to the U.S.-EU Leaders at the Summit itself (emphasis added).

In 2013, the year in which the EU member states instructed the European Commission to negotiate a new EU-US trade and investment agreement on their behalf, the TABD merged with the European-American Business Council (EABC)[19] to form the Trans-Atlantic Business Council (TABC), the major driving force behind TTIP. Over cocktail parties at government venues, large corporations discussed with public officials "in a pleasant and discreet environment" about ways that societies across the Atlantic could best serve the interests of these elites (Corporate Europe Observatory, 2014b). In contrast with their own deep involvement with the content of such trade and investment treaties, large corporations have stressed the importance of *not* taking time to consult non-governmental organizations (NGOs) and the public. At the U.S. Senate hearing on the TTIP, in order to gauge the necessity for passing the Trade Promotion Authority (or Fast Track) bill on the TTIP, Senator Thomas Carper asked business representatives to give him "the best argument... against Trade Promotion Authority... and then rebut it." Michael L. Ducker from Federal Express replied:

> I think that anybody would say the best argument against it would be the lack of collaboration and participation from large groups of people. And I would rebut that argument to say that, at the pace that commerce and trade deals are moving around the world, that we have to have speed to market in this case for U.S. business and U.S. trade.

Similarly, Ricks of Eli Lilly said, "I suppose the argument against it is to make sure all interests are well-represented, but I think, when one is negotiating, it is important to empower the people at the table to make the tradeoffs that are in the best interest of the country" (Committee on Finance, U.S. Senate, 2013: 16). Across the Atlantic, an internal Commission document listing tips for communicating the TTIP was leaked to a civil society group. In order to "reduce fears and avoid a mushrooming of doubts," the Commission called for a holistic public relations approach that involved the "management of stakeholders, social media and transparency" (Corporate Europe Observatory, 2013). Borrowing the Commission's own chilling concept, "managing transparency," as the title of his article, the British journalist and ecologist George Monbiot analyzed the panic spreading through the Commission. "Its plans to create a single market incorporating Europe and the United States, progressing so nicely when hardly anyone knew, have been blown wide open." All over Europe people were asking why they were not consulted; for whom it was being done (Monbiot, 2013).

In mid-December 2013 Michael Froman, the U.S. Trade Representative, and Karel De Gucht, the European Commissioner for Trade, received an open letter signed by some 200 NGOs, mainly from the U.S. and Europe but also from Latin America, Asia, and Africa, requesting the removal of the ISDS from the TTIP.[20] Under pressure, the European Commission agreed to consult the public on ISDS provisions in the TTIP. The result of the consultation, which the Commission announced in January 2015, was that, of the nearly 150,000 replies to the online consultation on the ISDS, about 97% opposed the inclusion of the ISDS in the TTIP (European Commission, 2015). Faced with overwhelming opposition to the ISDS, the Commission decided to give the transnationals-only judicial system a new name—the Investor Court System (ICS) and made some cosmetic changes to the old ISDS system.

Although the TTIP negotiations have been suspended,[21] neither European nor American societies are safe from corporate ISDS attacks[22] from the other side, owing to the signing of the CETA, which provisionally went into force in 2017. For multinationals operating in all three economies, ISDS attacks could be launched via Canadian subsidiaries, even if the SCAMD elites' attempt[23] to revive the TTIP negotiations fails. The obstacle for a full application of the CETA is the pending ratification by individual national and regional parliaments of EU member states. In October 2016 Belgium's Wallonia parliament rejected CETA, mainly because of its objection to the ISDS (even after being renamed and reformed as the ICS). While some saw the Wallonia parliament's rejection as an "heroic stand... for democracy," former EU Trade Commissioner Peter Mandelson described the veto as "a dagger at the heart of European trade policy" and argued that "it was time to end the scrutiny of trade treaties by national parliaments" (Olivet, 2016). In the end, the Wallonia government was pressurized into reneging on its veto. In exchange, the European Court of Justice will interpret whether the ICS provisions are legal under European law. For UN human rights expert Alfred de Zayas, such a development manifests "a culture of bullying and intimidation [which] becomes apparent when it comes to trade agreements that currently get priority over human rights." He warns that the corporate-driven CETA, negotiated outside public and parliamentary scrutiny, was compatible with neither the rule of law nor democracy and human rights (Tasch, 2016).

Among the reasons that the CETA is incompatible with human rights is the fact that "it goes in the opposite direction of our international commitments to limit global warming below a temperature rise of 2°C" (Hulot, Suzuki and Mayrand, 2016, cited in Corporate Europe Observatory, 2016: 5). Even with the ICS provisions pending and CETA only partly enforced, on the one-year anniversary of the Treaty, the Canadian government boasted the success of CETA by highlighting the increase in exports of the environmentally destructive tar sands oil to Europe. In fact, boosting tar sands exports to Europe from about 4,000 barrels per day in 2014 to about 725,000 barrels per day by 2020 was a main attraction of the CETA for Canada to begin with. While science says that "85% of [Canada's] 48 billion of barrels of bitumen reserves [must] remain unburnable" to meet the 2°C limit target (McGlade and Ekins, 2015: 190), Prime Minister Justin Trudeau insists that "no country would find 173 billion barrels of oil in the ground and just leave them there" (Patterson, 2018). The official website of Global Affairs Canada under his administration details "Opportunities and Benefits of CETA for Canada's Oil and Gas Exporters" and emphasizes that "the EU is the world's largest importer of oil and gas products, with imports totaling \$336 billion in 2016."[24] Against this background, further ICS protection to tar sands oil exporters could mean "game over for the planet" many times over.

Given its effectiveness in dispelling measures aimed at phasing out fossil fuels and rapid transition to renewables, ISDS is indispensable for industries that are faced with an existential threat such as fossil fuel companies that have the resources to use the ISDS and "no good reason not to" (Tienhaara, 2017: 241). To defend the ISDS, the EU has worked tirelessly in including the ICS not only in the CETA and the TTIP but in all its bilateral agreements with third countries.[25] More importantly, to protect and expand the ISDS, the EU has taken the initiative of establishing a Multilateral Investment Court (MIC) through the United Nations Commission on International Trade Law. The proposed MIC reinforces the one-sided nature of investment arbitration and upgrades ISDS from an *ad hoc* to a permanent arrangement that could systematically bypass domestic courts, further entrenching and institutionalizing the ISDS world-wide (Verbeek, 2018; CIEL and Seattle to Brussels Network, 2017; Rönnelid, 2018; Hoffmann, 2018). For anyone concerned with corporate power in blocking climate mitigation, the Commission's justification for actively promoting the MIC must be disturbing. Essentially, the EU considers that the bilateral ISDS, even after being renamed the ICS, is not stable and expansive enough. Hence, the MIC initiative "aims to replace existing bilateral mechanisms—including those in the over 1,400 investment treaties concluded by EU member states and other interested countries—with a permanent body to decide on international investment disputes." The MIC "should be for investment dispute settlement what the World Trade Organization is for trade dispute settlement, thus upholding a multilateral rules-based system" (European Commission, 2018).

The establishment of the MIC will complete the judicial system of the SCAMD structure, rounding up and disposing of the remnants of democracy that are still struggling to stand in the way of corporate interests. As a product of neocolonial power asymmetry, the ISDS should have lost its place in the 21st century. Instead, this profoundly unjust system has survived exactly as the neoliberal SCAMD elites planned, playing the role of the final arbitrator between the survival of climate-wrecking industries and that of the planet. Whether renamed as the ICS or expanded into the MIC, the goal of the ISDS remains to shackle democracy in locks and bolts, preventing, among other things, climate-mitigating regulations from being taken.

Notes

1 CPTPP was signed by 11 Asia-Pacific countries in 2018. Representing nearly 13.5% of global GDP, it is one of the largest free trade agreements.
2 Funding for the negotiations also came from the EC (Eberhardt et al., 2018: 32).
3 Later renamed the International Association of Oil and Gas Producers.
4 Today known as FuelsEurope.
5 *Public Citizen* has published an extensive study showing that terminating ISDS *has not* negatively affected countries' foreign direct investment inflows (Public Citizen, 2018).
6 Neither government nor civil-society groups such as labor unions or environment groups can bypass domestic courts and file lawsuits in tailor-made tribunals. The ISDS is indeed a unique privilege that exclusively serves the interests of big and transnational corporations.
7 In a review of 56 instances, 73% of the ISDS arbitrators took the expansive view that the FET could be interpreted autonomously from customary international law. In a further review of 137 instances, 83% of the ISDS arbitrators interpreted FET beyond its customary meaning and enlarged foreign investors' entitlements (Van Harten, 2016: 3).
8 This only includes disclosed corporate claims without counting the ones remaining in secret (Eberhardt et al., 2018: 7). For the distribution of treaties used for 706 ISDS lawsuits registered under World Bank's ICSID (International Centre for Settlement of Investment Disputes), see ICSID, 2019: 10.
9 Chevron acquired Texaco in 2001. By the time that the two companies merged to become Chevron Texaco, Texaco, which became a subsidiary of Chevron Corporation, had already performed oil operations in Ecuador for nearly four decades.
10 https://pcacases.com/web/sendAttach/2453.
11 Public Citizen has put together a list of egregious ISDS attacks. See www.citizen.org/sites/default/files/egregious-investor-state-attacks-case-studies_4.pdf.
12 According to James Hansen, "Canada's tar sands, deposits of sand saturated with bitumen, contain twice the amount of carbon dioxide emitted by global oil use in our entire history" (Hansen, 2012). In addition, it also has to be pumped through pipelines at higher pressures and hotter temperatures (Sierra Club, 2013: 1).

13 This tobacco carve-out is the result of ISDS lawsuits that the tobacco company Philip Morris brought against Uruguay and Australia after these countries adopted plain-packaging laws in accordance with World Health Organization recommendation to reduce smoking. Even though the case against Australia was thrown out for the technical reason that Philip Morris was not a Hong Kong company (the claim was made according to bilateral investment treaty between Hong Kong and Australia), Australian taxpayers will have to shoulder half of the $24 million legal fees, on top of substantial internal costs in the departments of health, trade, foreign affairs, and attorney generals. As to Uruguay, which could not afford to defend itself, it was the Bloomberg Foundation that funded the legal fees (Ranald, 2019).

14 Tienhaara also points out that, after decades of experiences of launching the ISDS over direct expropriations and tax increases, fossil fuel corporations are already comfortable with the ISDS. It is therefore fairly straightforward to apply this tactic to the new area of climate change policy and use the ISDS to challenge government decisions to ban fracking, scale back subsidies for dirty energies, shut down coal mines, or stop new oil and gas pipelines (2017: 241).

15 More than $1 billion in annual revenue, less than $10 billion.

16 More than $10 billion in annual revenue.

17 The Canadian company Lone Pine filed an ISDS lawsuit against its home country, Canada, via its U.S. subsidiary (Corporate Europe Observatory, 2014a). Similarly, the U.S. cigarette company Philip Morris restructured the company in order to use its Hong Kong subsidiary to bring an ISDS claim against Austria, with which Hong Kong has an investment treaty whereas the U.S. does not. See UNCTAD Investment Policy Hub website, available at http s://investmentpolicyhub.unctad.org/ISDS/Details/421.

18 www.transatlanticbusiness.org/tabd/.

19 Among the founding members of the EABC, Akzo Nobel, BASF, BP, ICI, and Philips were also members of the European Roundtable of Industrialists.

20 The letter is available at: http://corporateeurope.org/sites/default/files/attachm ents/ttip_investment_letter_final.pdf.

21 Awareness campaign by civil-society groups such as Public Citizen made TTIP negotiations difficult. See www.citizen.org/our-work/globalization-and-trade/ ttip-investment-map. In April 2015 an open letter from judges and economists further reduced the credibility of TTIP negotiations. See www.washingtonpost. com/r/2010-2019/WashingtonPost/2015/04/30/Editorial-Opinion/Graphics/oppo se_ISDS_Letter.pdf?tid=a_mcntx&noredirect=on.

22 Such attacks should be seen not as attacks on individual societies but as attacks on the planet, as long as they are related to climate mitigation.

23 In July 2018 European Commission President Jean-Claude Juncker visited President Trump and issued a joint statement which stated that the EU and the U.S. will strengthen strategic cooperation with respect to energy. Juncker promised to import more liquefied natural gas (LNG) from the U.S. to diversify its energy supply. http://europa.eu/rapid/press-release_STATEMENT-18-4687_en.htm.

24 https://www.international.gc.ca/trade-commerce/trade-agreements-accords-comm erciaux/agr-acc/ceta-aecg/business-entreprise/sectors-secteurs/OGE-EPPG.aspx?la ng=eng Last date modified 19 September 2017. Accessed 13 April 2019.

25 Including EU-Singapore, EU-Vietnam and EU-Mexico bilateral investment treaties. For all its ongoing bilateral negotiations with third countries, including Indonesia, Philippines, Myanmar, India, China, and Chile, the EU has also proposed to include the ICS (Friends of the Earth Europe, 2017: 2).

References

AFTINET. (2018). "ISDS Tribunal Says Ecuador Must Pay The Cost of Chevron's Pollution of Indigenous Communities." *Australian Fair Trade & Investment Network*. 13 September.

Bonnitcha, Jonathan, Lauge N. Skovgaard Poulsen, Michael Waibel. (2017). *The Political Economy of the Investment Treaty Regime*. Oxford: Oxford University Press.

Business & Human Rights Resource Centre. Texaco/Chevron Lawsuits (re Ecuador).

CETIM. (2015). "Chevron's Activities Impair Freedom of Expression of Victims, Academics, Students and Activists," 6 November.

CIEL and Seattle to Brussels Network. (2017). A World Court for Corporations. Rosa-Luxemburg Stiftung.

Citizens Trade Campaign. (2018). "State Legislators in All 50 States Want End to ISDS." 12 September.

Chan, Margaret. (2012). Galvanizing Global Action Towards a Tobacco-Free World. Keynote address at the 15th World Conference on Tobacco on Health. 20 March.

Chevron in Ecuador. (2010). Chevron's $27 Billion Liability In Ecuador "Glaringly Low" In light of BP Disaster. 8 June.

Coleman, Matthew, Lucinda A. Low, Steven K. Davidson, Jeffrey F. Pryce, Thomas Innes. (2014). "Foreign Investors' Options to Deal with Regulatory Changes in the Renewable Energy Sector," *Steptoe*. 23 September.

Committee on Finance, U.S. Senate. (2013). Hearing before the Committee on Finance, U.S. Senate, 113 Congress, first session. 30 October. 88–432-PDF.

Corporate Europe Observatory. (2013). Leaked European Commission PR Strategy: "Communicating on TTIP." 25 November.

Corporate Europe Observatory. (2014a). "Still Not Loving ISDS: 10 Reasons to Oppose Investors' Super-Rights in EU Trade Deals." 16 April.

Corporate Europe Observatory. (2014b). "TABC Invite for A Cocktail at the Bavarian Representation in Brussels." July.

Corporate Europe Observatory. (2016). "The Great CETA Swindle." November.

Deutsche Welle. (2019). U.S. Arbitration Court Rejects Germany's Plea in Nuclear Phase-out Compensation Case. 5 September.

Eberhardt, Pia, Cecilia Olivet and Lavinia Steinfort. (2018). One Treaty to Rule Them All—The Ever-Expanding Energy Charter Treaty and the Power It Gives Corporations to Halt the Energy Transition. Brussels/Amsterdam: Corporate Europe Observatory, Transnational Institute.

Energy Charter Secretariat. (2018). In Memoriam: Ruud Lubbers, Founder of the Energy Charter Process. 15 February.

European Commission. (1994). Press Release. MEMO-94–75. 17 December.

European Commission. (2015). Report Presented Today: Consultation on Investment Protection in EU-US Trade Talks. 13 January.

European Commission. (2018). Commission Welcomes Adoption of Negotiating Directives for A Multilateral Investment Court. 20 March.

European Council. (1990). The European Council, Dublin 25–26 June. Reproduced from the Bulletin of the European Communities, No. 6/1990.

European Parliament. (2015). European Parliament resolution of 14 October 2015 on "Towards a new international climate agreement in Paris" (2015/2112(INI)).

Friends of the Earth Europe. (2017). The Multilateral Investment Court Locking in ISDS.

Gosden, Emily. (2017). "Rockhopper Launches Arbitration Claim against Italy," *ISDS Platform*. 23 March.

Grandoni, Dino. (2018). "The Energy 202: Big Oil and Gas Companies are Winners in Trump's New Trade Deal," *Washington Post* Blogs. 3 October.

Hansen, James. (2012). "Game Over for the Planet." *The New York Times*. 9 May.

Hoffmann, Rhea Tamara. (2018). The Multilateral Investment Court: A Stumbling Block for Comprehensive and Sustainable Investment Law Reform. European Society of International Law, Conference Paper no. 10/2018.

Hulot, Nicolas, David Suzuki and Karel Mayrand. (2016). "Il faut saisir l'occasion de conclure un accord climato-compatible," *Le Devoir*. 13 October.

ICSID. (2019). The ICSID Caseload Statistics. International Center for Settlement of Investment Disputes.

Kendall, Clare. (2008). "Amazonian Chernobyl—Ecuador's Oil Environment Disaster," *The Telegraph*. 8 August.

Landau, Toby. (2014). ISDS: The Devil in the Trade Deal. Australian Broadcasting Corporation. 12 September.

McGlade, Christophe and Paul Ekins. (2015). "The Geographical Distribution of Fossil Fuels Unused When Limiting Global Warming to 2oC," *Nature*, Vol. 517, 187–203.

Mayer, Jane. (2011). "Taking It to The Street," *The New Yorker*. 20 November.

Monbiot, George. (2013). "Managing Transparency," *The Guardian*. 2 December.

Nelsen, Arthur. (2016). TTIP: Chevron Lobbied for Controversial Legal Right as "Environmental Deterrent," *The Guardian*. 26 April.

NRDC. (2015). Press Release. Leading Scientists, Economists Urge Rejection of Keystone XL Pipeline. February 11.

Oil Change International. (2013). "The Keystone XL pipeline will lead to a surplus of heavy crude oil on the Gulf Coast that will be exported."

Olivet, Cecilia. (2016). "Wallonia's Heroic Stand Against CETA Is A Stand for Democracy," *EUobserver*. 28 October.

Patterson, Brent. (2018). "CETA Anniversary Sees Crude Oil Exports Increase," *Rabble.CA Blog*. 17 September.

PSI. (2018). ISDS Reforms Shove Human Rights under Corporate Red Carpet. 11 September.

Ranald, Pat. (2019). "When Even Winning Is Losing. The Surprising Cost of Defeating Philip Morris over Plain Packaging," *ISDS Platform*.

Rönnelid, Love. (2018). An Evaluation of the Proposed Multilateral Investment Court System. BUE/NGL Research Report.

Ross, Alison. (2017). "Could Brexit Trigger Investment Claims?" *Global Arbitration Review*. 19 June.

Public Citizen. (2018). Termination of Bilateral Investment Treaties Has Not Negatively Affected Countries' Foreign Direct Investment Inflows. Research Brief. April.

Schultz, Thomas and Cédric Dupont. (2015). "Investment Arbitration: Promoting the Rule of Law or Over-Empowering Investors? A Quantitative Empirical Study," *The European Journal of International Law*. Vol. 24, no. 4, 1147–1168.

Sierra Club. (2013). Keystone XL Pipeline.

Stone Sweet, Alec and Florian Grisel. (2017). *The Evolution of International Arbitration*. Oxford: Oxford University Press.

Tasch, Barbara. (2016). UN Human Rights Expert: CETA is Incompatible with the Rule of Law, Democracy, and Human Rights. *Business Insider*. 29 October.

Telesur. (2018). The Permanent Court of Arbitration in The Hague Rules Against Ecuador, Favors Chevron. 7 September.

Tienhaara, Kyla. (2017). "Regulatory Chill in a Warming World: The Threat to Climate Policy Posed by Investor-State Dispute Settlement," *Transnational Environment Law*. 7(2): 229–250.

TransCanada. (2016a). Notice of Intent to Submit a Claim to Arbitration under Chapter 11 of the North American Free Trade Agreement.

TransCanada. (2016b). Request for Arbitration Under the Convention on the Settlement of Investment Dispute between States and Nationals of Other States and the Institution Rules and Arbitration Rules of the International Center for Settlement of Investment Disputes and Chapter 11 of the NAFTA.

Van der Vat, Dan. (2018). Ruud Lubbers (Obituary). *The Guardian*. 19 February.

Van Harte, Gus. (2015). An ISDS Carve-Out to Support Action on Climate Change. Osgoode Hall Law School Legal Studies Research Paper No. 38. Vol. 11(8).

Van Harte, Gus. (2016). Foreign Investor Protection and Climate Action: A New Price Tag for Urgent Policies. Osgoode Hall Law School Legal Studies Research Paper No. 21. Vol. 12(5).

Vattenfall. (2009). Vattenfall AB, Vattenfall Europe AG, Vattenfall Europe Generation AG v. Federal Republic of Germany. Request for Arbitration (ICSID Case No. ARB/09/6).

Verbeek, Bart-Jaap. (2018). "Same Old, Same Old: The EU Pushes ISDS 2.0.," The Centre for Research on Multinational Corporations (SOMO). 28 March.

7 Surviving Democracy

The pattern is clear: If climate mitigation takes a form and with a pace disapproved by the SCAMD elites, it simply cannot happen. Throughout the journey of its ascent, neoliberalism has kept democracy as its closest ally. Without the façade of being democratic and inclusive, neoliberals' exclusionary grasping power would have only limited space in society. By using democracy to dull consciousness, nurture complacency, and delay resistance, neoliberalism succeeded in turning liberal democracy into zombie democracy, with devastating consequences. To survive the ecological impacts caused by neoliberalized democracy, it is important to simultaneously identify the *root cause* of neoliberalization and address the problem of *time*. The fact that effective climate action is running out of time does not mean that time is on the side of neoliberals. Apart from the fact that climate change affects *everyone*, the double movement that Polanyi described has been unfolding for some time with the rise of fascism and other forms of popular resistance, such as the Occupy Movement. Increasingly, neoliberal grasping has to resort to explicit coercion with decreasing ability to tend to the manners of grasping as society comes to understand neoliberal public relations tactics, an indication that neoliberalism has run its course, with the election of Donald Trump as U.S. President embodying the last gasp of neoliberalism (Watkins and Seidelman, 2019). The question is whether the last straw for the neoliberal system will come before the last straw for the ecosystem as we know it. Movements and initiatives such as Extinction Rebellion and the Green New Deal are critical for providing vision and demanding immediate action. In order to add to the force trying to expedite the implementation of effective mitigation measures by clearing out neoliberal roadblocks, in this chapter I draw implications from the analysis of the book for the root causes of neoliberalization and formulate suggestions accordingly.

A significant part of the SCAMD elites' capacity in controlling the choke-holds of effective climate action comes from the blending of public and private roles. The agile interchangeability of identities between policymakers in the political domain and corporate elites in the market domain, whereby each aids the other in ensuring the smooth enclosure of society and commons, is a valuable lesson for society to learn. In the civic domain, citizens and voters direct their actions toward governments and politicians. In the market domain, investors and consumers can take initiatives for re-sequencing market activities (recalling Galbraith discussed in Chapter 4) and re-embedding market in society (recalling Polanyi discussed in Chapter 2). The interchangeability of identities among citizens and voters in the political realm and investors and consumers in the realm of the market is essential if we are to urgently clear obstacles to effective mitigation. In what follows I illustrate the point by describing how the fossil fuel divestment movement has helped to burst the carbon bubble. I then move on to discuss the meat bubble and its profound relationship with the core message of the book.

The fossil fuel divestment movement calls on institutions such as charities, churches, universities, foundations, and pension plans to remove from their portfolios stocks or bonds of fossil-fuel related companies.[1] Among the best-known groups involved in the movement are 350.org, Divest-Invest, and Fossil Free.[2] In a seminal article published in *Rolling Stone* magazine in 2012, Bill McKibben, the founder of 350.org, effectively imprinted three numbers in the minds of his readers: 2°C, 565 Gigatons, and 2,795 Gigatons. 2°C was the maximum increase in the average global temperature that world leaders agreed to commit to at the Copenhagen Climate Change Conference in 2009—a temperature level that scientists estimated would be fatal for many humans and non-humans across the planet. 565 Gigatons was the maximal amount of carbon dioxide that humans could still emit into the atmosphere (in 2012) before reaching the 2°C limit. 2,795 Gigatons was the amount of carbon already contained in the known coal, oil and gas reserves owned by the fossil-fuel companies and countries such as Venezuela and Kuwait. In short, it was the fossil fuel that we were planning to burn. While the world had not burned this fossil fuel yet, it already figured in share prices. Companies were borrowing money against it and nations basing their budgets on the presumed returns from their patrimony. These reserves were their primary asset, giving their companies their value. "It's why they've worked so hard these past years to figure out how to unlock the oil in Canada's tar sands, or how to drill mines beneath the sea, or how to frack the Appalachians" (McKibben, 2012).

The "[f]ossil-fuel industry... has become a rogue industry, reckless like no other force on Earth. It is Public Enemy Number One to the survival of our planetary civilization." The article soon became one of the most widely read pieces in the history of *Rolling Stone*, generating more online reads, comments, likes, shares, and Twitter mentions than any article the magazine had ever published. Together with a U.S. nationwide "Do-the-Math" fossil-free tour, the article helped to popularize the idea of fossil fuel divestment (Bigelow, 2013; Stephenson, 2012; Klein, 2015).

As of September 2018, a total of 985 institutional investors with endowments and portfolios worth $6.24 trillion in assets have committed to divesting from fossil fuels. Compared with 2014 when the committed assets were $52 billion, there has been a 11,900% increase. Consulting firm Arabella Advisors considers that fossil fuel divestment is now "a mainstream financial movement mobilizing trillions of dollars in support of the clean energy transition," creating "a negative material impact on the fossil fuel industry." Broader, divestment-related actions such as defunding[3] also "directly reduce fossil fuel emissions by slowing the expansion of the industry" (Arabella Advisors, 2018: 1–2). Shell's 2017 *Annual Report* corroborates this view, stating that if the fossil fuel divestment movement were to continue, it could have an adverse effect on the price of the company's securities and its ability to access equity capital markets. Accordingly, the ability of the company to use financing for future projects may be adversely affected—a situation that could also adversely affect potential partners' ability to finance their portion of costs, either through equity or debt (Shell, 2018: 13).

Similarly, finance experts noticed that "global stock indexes without fossil fuel holdings have outperformed otherwise identical indexes that include fossil fuel companies. Fossil fuel companies once led the economy and world stock markets. They now lag." "[T]he rationale for investing in it is untenable" (Sanzillo *et al.*, 2018: 2). Studies by Trinks *et al.* suggest that excluding fossil fuel stocks does not impair portfolio performance (2018), and Henriques and Sadorsky found that divesting from fossil fuels increased financial returns for investors (2017).

According to Schifeling *et al.*, however, the "ultimate effect" of the fossil fuel divestment movement "was not so much financial as on the terms of the debate over climate change." Sifting through 42,000 news articles about climate change between 2011 and 2015, they found that the divestment campaign had expanded rapidly as a topic in worldwide media. In the process, "it disrupted what had become a polarized debate and reframed the conflict by redrawing moral lines around acceptable behavior." The radical position of the movement had the effect of enhancing the viability of progressive issues, "creating opportunities for more moderate groups and issues to become more influential" (Schifeling *et al.*, 2017). It was this stigmatizing effect that got Ben van Beurden, the CEO of Shell, worried, stating "I do think trust has been eroded to the point where it starts to become a serious issue for our long term future" (Geman, 2017). In a study by Oxford University's Stranded Assets Programme, Ansar *et al.* highlighted that, in almost every divestment campaign they reviewed "from adult services to Darfur, from tobacco to South Africa, divestment campaigns were successful in lobbying for restrictive legislation affecting stigmatized firms" (Ansar *et al.*, 2013: 14). Considering the pattern, it is unlikely that fossil fuel divestment will "prove to be an exception… given that it is the fastest growing divestment movement in history" (Lenferna, 2018: 86). As a result of the fossil fuel divestment movement, previously marginalized policy options such as carbon tax regained some ground. The shift also "helped translate McKibben's radical position into new issues like 'stranded assets' and 'unburnable carbon'." Terms like these adopt the "language of financial analysis." When media such as *The Economist, Fortune,* and Bloomberg began to use them in their reports, these concepts became more legitimized within business circles, turning the battle cry of divestment into an alarm bell on financial risks. "By being addressed in these financial publications, the carriers of the message shifted from grassroots activists to investors, insurance companies and even the Governor of the Bank of England" (Schifeling *et al.*, 2017).

However, it is exactly the ability of the movement to engage with the market, i.e., sending signals to investors, that drew criticisms from some progressives. Parenti argues that a major flaw in the divestment campaign is that it ignores the powerful role of governments. "The only force on earth that can really control Exxon is the U.S. government." Yet instead of pressurizing the government to do the right thing, grassroots actions are now focused on divestment (Parenti, 2013). Jacobsen likewise criticized McKibben's belief in a market-based solution (2018: 24). In the same vein, Mayes *et al.* argue that, by leveraging free market mechanisms as the principal apparatus for social change, the fossil fuel divestment movement "endorses if not legitimates the reduction of state responsibility in favor of increased responsibility on the part of civil society." This "neoliberal mode of protest politics" fails to focus on getting governments to take action but rather places the responsibility of ensuring corporations behave responsibly on society in general and shareholders in particular. By framing responsibility as economic rationality and pragmatism, the movement also creates "fuzzy boundaries between moral and economic actors" and "marketizes social justice" (Mayes et al., 2017: 134, 140).

The arguments of these critics suggest that they do not see states and large corporations, particularly Big Oil, as having long blended into one entity, overseeing human affairs and allocating world resources from a position above both market and democracy. The critics' understanding of market and democracy as being more or less autonomous from one another and belonging to separate realms reflects how market and democracy used to and ought to work, but not how they actually do work. The neoliberal art of exclusion through inclusion has "worked" so marvelously only because government, on which Parenti, Jacobsen, and Mayes *et al.* believe activist energy should focus, has been so intermeshed with business. To rehabilitate government back into doing what it is supposed to do as these critics rightly suggest, the SCAMD structure has to be undermined with strategies that recognize the "oneness" of government and large corporations and the fact that both draw "talents" from the same neoliberal elite pool. Without flanking maneuvering with help from investors and consumers in the realm of the market, citizen protest in the realm of politics alone will not only take too long, but may also fail to rehabilitate governments who receive nutrition and oxygen mainly from corporations rather than citizens. The SCAMD elites might look down at civic disobedience with amusement or with annoyance, but never with the belief that they can be touched or that the structure needs to be changed. By reading statements from Big Oil, for instance, one cannot help but get the impression that its leaders not only believe that they rule from the top, but also that they see nothing wrong with that picture. In acknowledging the disappearing of societal acceptance of the fossil-fuels-dominated energy system, the regrets that van Beurden expressed were not with the way that his company has behaved, but with the irrationality of the public in handling the fact that the transition of the global energy system, which in his opinion was "now dominated by fossil fuels," was a decades-long endeavor. Because of the irrationality on the part of the public, it was very hard to get discussion "in the *right* spot." He acknowledged that the company was partly to blame "because we have *allowed* the discussion to drift into a *weird place*, and it is very difficult now to get it back to a more *rational* place." He lamented that emotional choices and attitudes were now an integral part of the lifestyle of the public, which was no longer driven by "common sense," something that "you can expect from a company like us" (Geman, 2017, emphasis added).

Precisely because corporations and governments are so confused about their proper roles, the so-called flaws in the divestment movement are in fact profound strengths. The movement recognizes and begins to address the existence of the SCAMD structure, emphasizing that the state *alone* cannot bring down Big Oil and fossil fuel companies. The divestment movement understands that by allowing companies to grow so large and powerful, the capitalist system has already rotted out the governmental system—a fact that explains why mainstream environmental campaigns' focus on government regulation in the past 30 or more years have failed to stop runaway emissions. "Divestment is an escalation precisely because government regulations have proved elusive, ineffective and even dangerous" (Bourqui, 2013). Likewise, the Gofossilfree campaign stated that the fossil fuel industry "has our political process in shackles with its financial might." The acts of divestment are thus aimed at taking back power from the fossil fuel industry and creating a mandate for political leaders (gofossilfree.org, cited in Mayes et al., 2017: 144).

Mayes *et al.* worry that by appealing to the logic of the market and emphasizing the importance of avoiding the "carbon bubble," campaigners and divestors are in fact assisting the market to function well. Hence, rather than breaking free from neoliberal shackles, the divestment movement "reproduce[s] neoliberalizing ideologies and practices…, [and] reinforces a neoliberal shift away from the state as the key corporate regulator" (Mayes et al., 2017: 144). Such a criticism suggests that overthrowing neoliberalism requires stigmatizing market mechanisms as a whole, rather than just the rigged and unjust market rules. Neither Polanyi's nor Keynes' nor Galbraith's theories justify such blind market-bashing. If neoliberalism can be seen first and foremost as a PR scheme, as this book suggests, to fall victim to market paranoia is to become mired in the virtual reality that neoliberal social scientists have helped to create. The virtual reality of the self-regulating market is not truly reality but rather is what neoliberals want us to believe. Neoliberals have usurped, twisted, and contaminated words and meanings pertaining to the market (Hudson, 2017) for the purpose of deception and exclusion; the proper counter measure is to reclaim and honestly redefine the market, rather than denying that the market has a proper role in society. As a result, the divestment movement can be seen as a step to reclaim the market, placing critical elements such as morality and transparent information back into the consideration of demand and supply.

Ending the fairytale that the market is always "self-regulating" is part of the responsibility of the movement. Had the "self-regulating" mechanism of the market—rather than planning and political power—been the sole force behind the economy, fossil fuel companies would never have gotten where they are, with the coercive power to hold the world hostage. As Galbraith emphasized, it was never the planning that was the problem. The problem had always been the dishonest discourse steadfastly denying the fact that money, power, and planning played a crucial role in the market economy. What divestment does is to let planning and civic power play their role in the market transparently and honestly. In other words, it re-embeds the market in society. Progressive critics of the divestment movement are right, however, about the groping power of neoliberalism. That is why constant caution against greenwash and consumerism must be an integral part of divestment as well as other mobilizations of climate mitigation. In sum, rather than replacing or sidelining civic grassroots efforts, divestment flanks and accelerates the effectiveness of such actions. Moral and ethical investment/consumption as a form of disobedience directed against the SCAMD is something that the mitigation endeavor cannot afford to dismiss. Rather than shying away from using the market as a critical venue for revolt, the case for doubling down on this approach is strong.

An important source of inspiration for Bill McKibben in writing the 2012 *Rolling Stone* article came from reports from an independent financial think tank, Carbon Tracker, that factors in the impact of carbon emissions in its risk assessments of investment behavior. While studies done by Carbon Tracker have players in the financial markets as their targeted audience, it was these reports that gave rise to McKibben's idea to frame the issue the way that he did, driving home the information that fossil fuel companies have five times more carbon in stock under the ground than can be burned. In a similar vein, the Farm Animal Investment Risk and Return initiative (FAIRR) works to raise awareness about risks and opportunities linked to a climate-related "meat bubble." Noting that it is becoming clear that meat is "on a similar pathway to tobacco, carbon and sugar," FAIRR advised investors to "take action before further consumer pressure and regulation moves markets" (FAIRR, 2017: 26). The existence of a meat bubble alongside the carbon bubble represents an opportunity for SCAMD dissidents to double down on revolt through economic behavior.

The claim that the meat industry is in trouble is strongly backed by the Intergovernmental Panel on Climate Change (IPCC) of the United Nations (UN). In its special report, *Global Warming of 1.5°C*, which was published in October 2018, the IPCC considered reduction of meat consumption a key element in the materialization of the 1.5°C target: "There is increasing agreement that overall emissions from food systems could be reduced by targeting the demand for meat and other livestock products."

> Livestock are responsible for more GHG emissions than all other food sources. Emissions are caused by feed production, enteric fermentation, animal waste, land-use change and livestock transport and processing. Some estimates indicate that livestock supply chains could account for 7.1GtCo$_2$, equivalent to 14.5% of global anthropogenic greenhouse gas emissions (Gerber *et al.,* 2013). Cattle (beef, milk) are responsible for about two-thirds of that total, largely due to methane emissions resulting from rumen fermentation (Gerber *et al.*, 2013; Opio *et al.*, 2013). (IPCC, 2018: 197–8)

Despite ongoing gains in livestock productivity and volumes, the report made clear that such gains are limited in terms of climate mitigation "because of inefficiencies in the conversion of agricultural primary production (e.g., crops) in the feed-animal products pathway" (Alexander *et al.*, 2017). The report notes that science already knows that dietary shifts could contribute "one-fifth of the mitigation needed to hold warming below 2°C, with one-quarter of low-cost options" (Griscom *et al.*, 2017); it is just not yet clear how that role changes when the target is moved to 1.5°C (IPCC, 2018: 198).

Coinciding with the publication of the IPCC special report (2018) is an article published in *Nature* which, according to the lead author, Marco Springmann, is the first comprehensive study on the impact on the planet of global food production. The authors of the article estimate that in 2010 the food system emitted about the equivalent of 5.2 billion tons of carbon dioxide in greenhouse gas (GHG) emissions in the form of methane and nitrous oxide. Within the system, the production of animal products generated 72–78% of total agricultural emissions. The relative contribution of animal products to GHG emissions in agriculture will increase by 7–16% in 2050, when the GHG emissions from the food system as a whole will have jumped by roughly 87% from the 2010 baseline. Both the increase of overall emissions from the system and the relative contribution of animal products are mainly due to population and income growth (Springmann, Clark *et al.*, 2018: 2). The authors estimate that dietary changes towards more plant-based (flexitarian) diets could reduce GHG emissions by 56% (Springmann, Clark et al., 2018: 3). It shows that staying within the GHG boundary that allows a safe operating space for humanity "*requires ambitious dietary change towards more plant-based, flexitarian diets*" (Springmann, Clark *et al.*, 2018:4, emphasis added).

As striking as the findings of the heavy carbon footprint associated with animal product consumption are in studies such as this, these estimations, by expressing methane and nitrous oxide emissions in terms of "carbon dioxide equivalency," are likely to have severely *underestimated* the significant role that a dietary shift away from animal products can play in meeting reduction targets in the extremely short period of time—or a decade, using the assessment of the latest IPCC report. According to Wikipedia, "carbon dioxide equivalency" is "a quantity that describes, for a given mixture and amount of greenhouse gas, the amount of CO_2 that would have the same global warming potential (GWP), when measured over a specified timescale (generally, 100 years)." Given that the GWP of methane over 100 years is 34 times that of carbon dioxide, emissions of 1 million metric tonnes of methane is deemed to be equivalent to emissions of 34 million metric tonnes of carbon dioxide.

The 100-year timescale used in the standard global warming metrics spreads the heat-trapping potency of methane—a major by-product of the livestock industry—evenly over a period of 100 years, when methane stays in the atmosphere for only a decade and has the heat-trapping potency of about 100 times that of carbon dioxide when first released from its source. If humanity has a decade to react to the dire climate situation, sticking to the 100-year timescale is inadequate, given how much it skews the short-term heat-trapping impact of GHGs such as methane. The IPCC emphasized in its first assessment report that "relative forcings based on emissions [as opposed to relative forcings based on concentrations]... require a careful consideration of the radiative properties of the gases, their lifetimes and their indirect effects on greenhouse gases." Emphasizing that "there is no universally accepted methodology for combining all the relevant factors into a single global warming potential for greenhouse gas emissions..., a simple approach has been adopted here to illustrate the difficulties inherent in the concept" (IPCC, 1990: 58). The usefulness of "carbon dioxide equivalency" underpinned by GWP for illustration is therefore gained at the expense of "a careful consideration of the radiative properties of the gases, their lifetimes and their indirect effects on greenhouse gases." As a result, the radiative forcing stabilization level that needs to be achieved much sooner than 100 years is destined to overshoot, owing to the underestimation of the short-term impact of methane. "These errors become increasingly concerning over time as climate thresholds are approached" (Edwards and Trancik, 2014: 347).

The inclusion of the time factor as a reason for the world to become worried about the emissions of methane goes beyond the inadequacy of spreading the impact of an extremely potent but short-lived warming GHG evenly across a 100-year period in scientific calculations. The underestimation of the short-term impact of methane that occurred owing to the reliance on a notion designed for easier illustration through simplification is compounded by the long-lasting sea-level effect of this short-lived GHG. Even if man-made methane emissions stop today, resulting in zero anthropogenic methane in the atmosphere 10 years from now, given the amount of heat trapped by this powerful form of GHG during its lifespan, the sea-level rise through thermal expansion will continue for several centuries afterward. The persistence of the warming is due to "ocean thermal inertia: when GHG emissions stop the system is still not equilibrated with the peak radiative forcing and the ocean continues to take up heat." As a result, "the climate impacts of short-lived GHGs [such as methane] are far longer-lasting than would be implied by their atmospheric lifetimes alone." This finding further "heightens the importance of earlier mitigation actions" (Zickfeld *et al.*, 2017: 659, 662).

Global methane from in situ observations reached a new high in 2017 (WMO, 2018: 3). Methane is the second largest contributor to global warming after carbon dioxide; contributing about 17% of the radiative forcing by GHGs. Approximately 40% of global methane emissions are from natural sources; the other 60% are from anthropogenic sources. The anthropogenic sources have led to a 257% increase in atmospheric methane compared with the pre-industrial level (Riahi *et al.*, 2017: 161; WMO, 2018: 3). According to a 2006 study by the Food and Agriculture Organization of the UN, 35–40% of global anthropogenic emissions came from enteric fermentation and manure in the livestock industry (FAO, 2006: 112). Up-to-date estimation of livestock's contribution to global methane emissions will be much higher owing to population growth, increase in food demand, and changes in animal body mass, feed quality and quantity, milk productivity, and management of animals and manure (Riahi et al., 2017:161; Wolf *et al.*, 2017: 1).

The findings on the underestimation of the short-term impact of methane owing to standard "carbon dioxide equivalency" expediency and the huge time lag between the cease of emissions and the cease of methane-caused sea level rises, combined with the findings about the huge and rapidly increasing role that livestock plays in anthropogenic methane emissions, lead to a very clear conclusion: Immediate and radical reduction in anthropogenic methane emissions is critical; the consumption of meat, dairy, and eggs is climate recklessness.

The climate recklessness of the consumption of meat, dairy, and eggs, however, is far from confined to its contribution to methane emissions. Methane is not the only important GHG emitted from enteric fermentation and manure. Nitrous oxide, which is released not only from enteric fermentation and manure but also from chemical fertilizers applied to feed crops, is 298 times more potent in trapping heat than carbon dioxide on a 100-year timescale, as well as being an ozone destroyer (FAO, 2009: 64; Grace and Barton, 2014). To put in context the magnitude of nitrous-oxide-emitting chemical fertilizers used for sustaining animal consumption, it is important to point to the fact that, globally, livestock is the largest user of land resources. Approximately 80% of all agricultural land is dedicated to feed crops or grazing. One-third of total cropland is used for animal feed while 26% of Earth's ice-free terrestrial surface is occupied by pasture.

Altogether, 45% of the global surface area is dedicated to livestock (FAO, 2009: 54; Thorton *et al.*, 2011). The shocking demand for land in livestock industry further explains another major component of carbon emissions by the industry, namely the destruction of carbon-absorbing forests. Approximately 80% of deforestation in the Amazon is associated with livestock (Veiga et al., 2003; Alix-Garcia and Gibbs, 2017: 201).

As neoliberals would want us to believe, "consumer demand" or "individual choice" is what is stopping the phasing-out of climate-wrecking animal consumption. This, however, is an explanation that has taken the phenomenon of animal consumption completely out of the neoliberal context. Once the phenomenon is placed back in the neoliberal context, it becomes clear that, rather than being about the taste buds of consumers, animal consumption is, unsurprisingly, yet again about PR works, lies, and manipulation. An important but easily neglected fact that the study by Springmann, Clark *et al.* contains is that global and national dietary guidelines are not only incompatible with a healthy environment; they are also not aligned with the health benefits for people (Springmann, Clark et al., 2018: 3,5), the sole purpose such guidelines are supposed to serve. The comparison in the study by Springmann, Clark *et al.* is between "diets in line with global dietary guidelines for the consumption of red meat, sugar, fruits and vegetables, and total energy intake" on the one hand and "more plant-based (flexitarian) diets that more comprehensively reflect the current evidence on healthy eating by including lower amounts of red and other meats and greater amounts of fruits, vegetables, nuts and legumes" on the other, with the latter estimated to be able to reduce GHG emissions by 56% (Springmann, Clark et al., 2018: 3). The piece of information regarding the discrepancy between dietary guidelines and health benefits reveals the interlinked, mutually reinforcing relations among different actors within the SCAMD elite structure and the complex overall damage that this network does to the general public and the planet.

In the U.S., the forerunner to the Dietary Guidelines for Americans, published every five years by US Department of Health and Human Services (HHS) and the US Department of Agriculture (USDA), was the 1977 "Dietary Goals for the United States," a general food advisory for the public. A key figure who helped to draft the report was a nutritionist at Harvard, D. Mark Hegsted (Pearce, 2009). The first Dietary Guidelines for Americans, which appeared in 1980, advised the public to reduce total fat, saturated fat, and dietary cholesterol in their diet in order to prevent coronary heart disease. Omitted from the list was sugar, a dietary factor that scientists started to identify as linked to coronary heart disease in the 1960s. Hegsted's role in this omission was suggested by a study published by *JAMA Internal Medicine* in 2016 which found that Hegsted was paid by the sugar industry, starting in 1965, to conduct and publish research aimed at diverting public discourse and government decisions away from scientific findings revealing the negative role of sugar in health. The pattern of food industries treating the Dietary Guidelines, which have a direct impact on school meals and food labels for the five years that follow, as a sales channel has persisted. As a result, the number of Americans having type-2 diabetes or being obese has more than doubled since the Guidelines first appeared, with two-thirds of Americans currently being either overweight or obese (Kearns *et al.*, 2016).

The latest Guidelines for the period 2015–20 have seen the meat industry instructing the Federal government to strike down the Advisory Committee's recommendation—reached after reviewing thousands of studies—on eating less red meat. In contrast to the Committee's scientific opinion, the guidelines list red meat alongside seafood and poultry as elements of a "healthy eating pattern," despite findings linking red meat with cardiovascular disease and cancer.[4] The powerful trade group, the North American Meat Institute, representing companies such as Tyson Foods Inc., issued a statement celebrating the "commonsense policy document that all Americans can use to help them to make healthy food choices" (Evich, 2016). New York University nutritionist Marion Nestle, who participated in crafting past guidelines, said "I was told we could never say 'eat less meat' because USDA would not allow it" (Heid, 2016). In the first three quarters of 2015 the National Pork Producers Council spent $780,000 on lobbying; the National Cattlemen's Beef Association spent more than $112,000, with the Dietary Guidelines listed as a major target. The North American Meat Institute spent more than $220,000 on lobbying. What the money bought was the promises of heads of the USDA and HHS to steer clear of the "eat less meat" expert recommendations (Evich, 2016). The title of a recent book by Gerardo Otero says it all: *The Neoliberal Diet—Healthy Profits, Unhealthy People*. Human bodies are turned into refinery machines, with raw materials—meat, dairy, sugar, fat, chemicals, etc.—shoveled down throats and harmful ingredients absorbed by organs so that the value can be extracted by corporations in the form of profit. Instead of safeguarding the health of citizens, the state tends to dutifully fulfill its SCAMD role of maintaining the neoliberal order.

The SCAMD elites clearly will place neither the health of the public nor the 1.5°C-consistent emissions before the interests of large corporations. The link between a corporate attack on the health of people and that of the planet increases the effectiveness of using the market by enlightened citizens, consumers, and investors as the pathway for re-embedding the market in society. Progress made on one front increases the likelihood of victory on the other. An unhealthy population is good business for Big Pharma but a heavy burden on the atmosphere. In the U.S., the healthcare system contributed 9.8% of the national GHG emissions in 2013. If the U.S. healthcare system were a separate country, it would be the world's 13th-largest emitter of GHG, ahead of the entire U.K. (Eckelman and Sherman, 2016). Given the scientifically proven link between meat and cardiovascular disease and cancer, a plant-based diet will translate into billions of treatments and surgeries not having to be done, medicines not having to be prescribed, and supplies and equipment never having to be produced and shipped.[5]

Unlike the phasing-out of fossil fuel, which is dependent on government infrastructure from energy supplies to public transportation, the dietary shift away from meat and dairy products can easily be adopted at the individual level. Whereas divestment movements—whether from fossil fuel or meat, both being imperative—require a high level of organization and mobilization given the central role played by institutional assets/portfolio, ordinary consumers get to play a much more prominent and direct role in the case of plant-based shift in diet, owing to their greater autonomy over food choices in comparison with energy choices. As this book has demonstrated, as it stands, the state and big corporations are mutually constitutive in neoliberalized democracies. To create the possibility for the state to be weaned off big corporations and vice versa, enlightened citizens acting with interchangeable roles as consumers and investors are critical. Under this strategy, the fossil fuel industry and the meat industry are the entry points that make the most sense under the mitigation time constraint.

Measured against the core argument of this book, however, even a fast and dramatic global reduction in meat consumption that toes the fine line of killing, say, 2 billion rather than 2.1 billion animals a day in order not to exceed the emissions quota, is a strategy completely failing to address the root cause of climate change. Such actions, while qualifying as a technical fix, do not amount to a systemic overhaul that addresses the fundamental reason lying at the heart of the climate crisis. This book has taken exclusion and the excuses and PR works designed to justify or hide the act of exclusion as the core essence of neoliberalism. Climate change is just one of many—often interacting—residues of neoliberalism. Animal consumption is one other such residue, which has interacted with the environment, resulting in the meat bubble.[6] In the following section I explain why, apart from in extremely rare cases, endorsing animal consumption is an obedient display of endorsement of neoliberalism and exclusion.

Roughly speaking, humans consume animals for two reasons. They either believe that animals cannot feel pain and fear or do not consider it relevant whether animals are sentient. This overly simplified depiction of human behavior wrongfully assumes individual autonomy, which the book has highlighted is severely eroded under neoliberalism. The indoctrinated mantra, "free to choose," protects "not the individuals' right to buy," but "the seller's right to manage the individual" (Galbraith, 2007 [1967]: 270). The neoliberal art of exclusion through inclusion has at its disposal both enhancer and inhibitor. To trick the brains of individuals into deciding that dubious products are not really that harmful, for instance, neoliberals administer pleasure-enhancement to its subjects/victims. In the case of animal consumption, not only have neoliberals injected pleasure-enhancement into our brains via advertisement, pop culture, and mass media, but large doses of inhibitors have also been prescribed to ensure that empathy and compassion remain deeply dormant. Paul McCartney, a former member of The Beatles, famously said that if slaughterhouses had glass walls, everyone would be vegetarian. The truth is, as is the case with climate change, that if slaughterhouses had glass walls, legislation would have been in place banning animal consumption. Because of neoliberalism, not only is there no such legislation, but measures are also taken to ensure that slaughterhouses—as well as factory farms—are protected behind fortresses to prevent public scrutiny. Parallel to the phenomenon where carbon emissions drastically went up as more scientific knowledge about climate change surfaced, the drastic increase of animal consumption coincided with a radical increase of scientific findings about how smart, sentient, social, affectionate, sophisticated, rational, and even moral animals are. As scientific findings about animal sentience and beyond were published, including books such as *Good Natured: The Origins of Right and Wrong in Humans and Other Animals* (de Waal, 1997), *Pleasurable Kingdom: Animals and the Nature of Feeling Good* (Balcombe, 2006), *The Emotional Lives of Animals: A Leading Scientist Explores Animal Joy, Sorrow, and Empathy* (Bekoff, 2008), *Second Nature: The Inner Lives of Animals* (Balcombe, 2011), *Beyond Words: What Animals Think and Feel* (Safina, 2015), *The Soul of an Octopus: A Surprising Exploration into the Wonder of Consciousness* (Montgomery, 2016), *What A Fish Knows* (Balcombe, 2017), *Are We Smart Enough to Know How Smart Animals Are?* (de Waal, 2017), and *Mama's Last Hug: Animal Emotions and What They Tell Us About Ourselves* (de Waal, 2019), not to mention countless articles in scientific journals, individuals' participation in daily and routine commercial handling of animals through consumption increased, with animals slaughtered just for food reaching 3 billion every day.[7] The kind of "standard operations" is documented in the film Earthlings, which is listed under the genre of "horror" in IMDb (Internet Movie Database), even though it only documents what societies do, day in day out. Here, the narrator explains what goes on in a scene where animals are suffering painful and slow death, which nearly all of us contribute to through our consumption behavior:

> Death by anal electrocution is a crude process that requires a probe to be inserted in the rectum while the animal bites down on a metal conductor. Often times this inept procedure must be repeated to actually kill the animal. And the skinned carcasses seen here will later be ground up and fed to the animals still caged. (Earthlings, 2005)

"Animal welfare," i.e., making the way in which animals are raised and killed more "humane," is often considered the most reasonable solution for the inconvenient cruelty problem. Such a solution, however, is as reasonable as counting on fossil fuel companies to provide "clean fossil fuels" to solve the problem of climate change. As has been the case in climate mitigation, government plays an active role in neutralizing animal welfare requirements by protecting industries from having to make costly changes. In the U.S., the animal industry drafted the "Animal and Ecological Terrorism Act" model bill in 2002, which proposed placing anybody "entering an animal or research facility to take pictures by photograph, video camera, or other means with the intent to.... defame the facility or its owner" on a "terrorist registry." Since the purpose of the Act is to silence whistleblowers, bills modeled on the Act are known as "Ag-Gag" bills. Ag-Gag bills have proliferated in the U.S. since the 1990s, with 27 State Legislatures having sought to pass Ag-Gag laws thus far wherein nine succeeded, 16 were defeated, and two ruled unconstitutional (Gibbons, 2017). The development already created a chilling effect on undercover investigations of treatments of animals in factory farming, which is the only possible way for abusive treatments to come to light for the public (Heuer, 2016).

In May 2018 animal rights activists were charged with felonies by a Utah prosecutor and face prison terms of up to 10 years. During an undercover investigation, the animal activists found tens of thousands of turkeys "crammed inside filthy industrial barns, virtually on top of one another." The birds were suffering from "diseases, infections, open wounds, and injuries sustained by pecking and trampling one another." They were "barely able to stand, or were lying in their own waste, close to death." There were animals that "looked dead but were still breathing." "Hogwash" is as common in animal welfare issues as "greenwash" is in environmental issues. The turkey farm under the undercover investigation is a supplier to a company marketing itself as selling mountain-grown turkeys treated humanely. The activists do not see the case as one of the government failing to do its job of stopping abuse, violence, and unethical behavior. Instead, the government is almost entirely captured and "in active collaboration with the industry that commits these crimes, forcing citizens to expose them" (Greenwald, 2018). Meat industry lobbyists have long tried to "educate" the public about undercover videos showing content that "may seem troubling to someone unfamiliar with farming." Comparing watching animal cruelty to "seeing open-heart surgery for the first time," a lobbyist stated, "they could be performing a perfect procedure, but you would consider it abhorrent that they were cutting a person open" (Oppel, 2013).

The animal welfare "solution" is problematic not only because enforcement is difficult. At a fundamental level, it rejects the notion of basic rights and equality, calling into question the legitimacy of all equality-seeking movements, including "Occupy," "#me too," "black lives matter," and most importantly, climate action. The SCAMD elites described in this book serve as a mirror for defenders of animal consumption. Parallel to the phrase "capitalism for the poor, socialism for the rich" commonly used to describe neoliberalized societies, the phrase "utilitarianism for animals, Kantianism for people" (Nozick, 1974: 39)[8] is frequently used to criticize justification of "necessary" animal sacrifices. As this book has demonstrated, "Kantianism for people" is a mirage under neoliberalism, which is through and through utilitarian in nature. The consequence of fake "Kantianism for people" combined with "utilitarianism for animals" is hell for animals and humans in the lowest stratification of society, including factory farm- and slaughterhouse workers.[9] From a neoliberal perspective, it matters little whether the excluded are human or non-human, as long as its tactics work to train society to accept exclusion. Animal welfare narratives, by giving consumers peace of mind, enhance the sophistication of exclusion and circumvent the voices of animal rights in the same way that Big Oil's "clean and affordable energy" rhetoric helped to circumvent voices demanding fossil-fuel-free society.[10] Just as the SCAMD elites have taken their entitlement to tap into natural and public resources for their private profit for granted, the notion that, ultimately, animals are there to serve humans has become deeply engrained in society. Nussbaum, for instance, argues, "[i]f animals were really killed in a painless fashion, after a healthy and a free-ranging life, what then? It seems unclear that the balance of considerations supports a complete ban on killings for food" (Nussbaum, 2004: 315). Schinkel points out the "simple and seemingly innocent word *'after'*... is a treacherous word." "It sounds almost pleasant—to be killed in a painless fashion after a healthy life. If only it were true that the killing occurred *after* the animal's life—in fact, of course, the killing occurs *in the middle of* a healthy life" (Schinkel, 2008: 51). Like SCAMD elites who consider the role of allocating resources befittingly theirs, defenders of animal consumption exercise the justification by looking down on animals from a high-up position.

A closer look at the relationship between "sentience" and "life" not only reveals that the argument that "it is acceptable to take one's life *after* he, she, or it has lived a good life" fails to make sense,[11] but also offers further insights into neoliberalism's destructive power. Sentience serves the critical function of detecting danger and enhancing the chance of survival. It is only a means to the end of life, which has intrinsic value. As a result, animal welfarists' downgrading of life to be only secondary to sentience is preposterous and can be understood only as an excuse of the powerful for justifying the exploitation of the powerless. Planet Earth is at the brink of extinction precisely because neoliberalism has excelled at dulling the ability of society to detect danger. With tactics such as Buchanan's "methodological individualism" that herded the public to focus on the maximization of "individual utility function," society has been administered with high dosages of empathy- and compassion-inhibitors, losing its ability to adjust collective behavior even when a threat with the magnitude of climate change presented itself.

At the same time that neoliberalism dulled society's ability to feel, sympathize, and perceive danger, it ensured that non-sentient and money-sucking entities like large corporations were granted basic rights. In *Citizens United v. Federal Election Commission*,[12] large corporations were seen as equivalent to humans, entitled to the protection of the basic right of freedom of speech, expressed in the form of campaign-contribution. The outrage that the ruling stirred has profound implications not only for the discussion of neoliberalism, but also for that of animal rights. The anger emanates from the ruling's prioritization of non-conscious and non-sentient entities—large corporations—over conscious and sentient citizens as participants of democracy when it was the latter rather than the former that are the ones vulnerable to the deprivation of inalienable natural rights. That beings that are conscious and sentient *should* have inalienable basic rights is an inescapable conclusion from the condemnation of the neoliberal landmark ruling. Like all other basic rights, however, the recognition of animal rights is inconvenient to those who have an interest in violating them. Although animal rights are often dealt with as a philosophical question, they are really only a simple matter of naked power.[13] This discourse-reality discrepancy is not unique for animals, but also characterizes debates concerning human rights. The real contention in most rights discussions is often not about whether a group or a class of people—i.e., people living in climate-sensitive regions, the poor, women, people with a sexual orientation different from the dominant group, or people with skin color different from the dominant group—*has* positive rights, but about whether the rights of those who have an interest in inflicting—and who are inflicting—harm on the vulnerable group can be taken away. In this light, animal welfare arguments are in line with the neoliberal logic that seeks sophisticated narratives for justifying the behavior of the powerful. They protect the perpetrators of mistreatment against animals rather than animals themselves.

The "great thinker" who made animal welfare arguments seem "ethical" was the 18th–19th-century British philosopher Jeremy Bentham. Peter Singer saw Bentham as a pioneer in making the basic moral principle of equal consideration of interests applicable to non-human species. For Bentham, as animals have the capacity for suffering, their interests deserve equal consideration (Singer, 2009 [1975]: 6–7). But what is Bentham's concept of "equal consideration of interests?" He famously said, "It is the greatest happiness of the greatest number that is the measure of right and wrong." To understand what Bentham really meant, it is necessary to know who he was and what he stood for, which neatly brings animals, climate, and neoliberalism together. Bentham featured prominently in Polanyi's discussion of enclosure in *The Great Transformation*. According to Polanyi, Bentham "despised equalitarianism, ridiculed the rights of man and bent heavily toward laissez-faire" (Polanyi, 2001 [1944]: 115). It is within the context of a strong objection against government intervention in helping the poor that his notion of utility maximization must be understood.

> The calculus of pain and pleasure required that no avoidable pain should be inflicted. If hunger would do the job, no other penalty was needed. To the question, "What can the law do relative to subsistence?" Bentham answered, "Nothing, directly." Poverty was Nature surviving

in society; its physical sanction was hunger. "The force of the physical sanction being sufficient, the employment of the political sanction would be superfluous." All that was needed was the "scientific and economical" treatment of the poor. (Polanyi, 2001 [1944]: 122)

In short, Bentham's idea of utility maximization concerned the "acceptance of near-indigence of the mass of the citizens as the price to be paid for the highest stage of prosperity" (Polanyi, 2001 [1944]: 123). Bentham not only *talked* about the "scientific and economical" treatment of the poor, he *acted* upon it. Bentham's Panopticon not only made supervising inmates in jail cheap and effective, it was also turned into convict-run factory, with the place of the convicts taken by the poor. The poor-turned-inmates were further turned into free labor for Bentham and his brother's private business venture. This model further "merged into a general scheme of solving the social problem as a whole," with "surplus from the labor of the unemployed... turned over to the shareholders" (Polanyi, 2001 [1944]: 111, 115).

His Industry-Houses, on the Panopticon plan—five stories in twelve sectors—for the exploitation of the labor of the assisted poor were to be ruled by a central board set up in the capital and modeled on the Bank of England's board, all members with shares worth five or ten pounds having a vote. (Polanyi, 2001 [1944]: 111)

Today's neoliberalized societies look eerily similar to Bentham's design. Beyond the apparent similarities of mass incarceration and exploitation of inmates (particularly in the U.S.) with Bentham's Panopticon (Hedges, 2013), there is also the intensification of state surveillance, the criminalization of dissent, and the "extended panoptic mechanisms" that the U.S. government has deployed to "keep friends and enemies alike under a condition of constant surveillance" (Gill, 2015). The deeming of a system where the rich few control and exploit the rest of society as "good" because the overall "utility" has been maximized[14] is likewise Benthamite. Many of the neoliberal signature ideas can be traced to Bentham, who "was the first to recognize that inflation and deflation were interventions with the right of property: the former a tax on, the latter an interference with, business" (Polanyi, 2001 [1944]: 234). The exclusionary nature shared by Bentham and neoliberalism is striking. Like neoliberalism, it was freedom with property that formed the essential part of Bentham's conception of individual liberty. "The condition most favorable to the prosperity of agriculture exists, when there are no entails, no unalienable endowments, no common lands, no right or redemptions, no tithes..." (Polanyi, 2001 [1944]: 189).

Bentham's idea of right and wrong sits comfortably with that of Hayek, Buchanan, SCAMD elites, and particularly the big polluters. Bentham's theory is detrimental to animals, humans, and to the climate. This book has highlighted that burning fossil fuels is *not* the root cause of climate change. Rather, the root cause is exclusion and exploitation successfully disguised by the neoliberal art of exclusion through inclusion. Climate change is only one residue of such exclusion and manipulation; animal consumption is another. Whether sugarcoated with the label of "utilitarianism," "public choice theory," or "rational choice theory," these tools used by and serving the interests of those sitting at the top, having been exposed, no longer deserve a place in the climate emergency. Climate mitigation is not for some; it is for all. Nobody deservedly serves just a functional purpose for others. The disappearance of polar bears is bad not because *humans* can no longer marvel at them. Mass extinction is bad not because it affects food supply for *humans*. Utilitarianism, neoliberalism, anthropocentrism, and climate disaster have the same core of exclusion. The only human–animal relationship compatible with inclusive societies that rejects the SCAMD structure and any other forms of exploitation, inequality, and exclusion is the one suggested by Regan, Francione, Donaldson, and Kymlicka. For Regan, animals, just like us, "have certain basic moral rights, including in particular the fundamental right to be treated with the respect that, as possessors of inherent value, they are due as a matter of strict justice." Therefore, like us, "they must never be treated as mere receptacles of intrinsic values (e.g., pleasure)" (Regan, 2004 [1983]: 329). For Francione, animals can never be protected from human harm unless the legal status of animals as property is abolished, as "When the legal system mixes rights considerations with utilitarian considerations and only one of two affected parties has rights, then the outcome is almost certain to be determined in favor of the right holder" (Francione, 1995: 107). It is therefore necessary to abolish the property status of animals. Donaldson and Kymlicka further apply citizenship theory to animal rights to help to envision a just world, where "animals in the wild vulnerable to human invasion and colonization" should be seen as "forming separate sovereign communities on their own territories;" "liminal opportunistic animals" are akin to "migrants or denizens who choose to move into areas of human habitation;" and "domesticated animals" should be seen as "full citizens of the polity because of the way that they have been bred over generations for interdependence with humans" (Donaldson and Kymlicka, 2011: 14).

The repairing of human–animal relations has a critical role to play in undoing Buchanan's individualization of organismic society (see chapter 3: 35–6) and revitalizing democracy, which is essential for climate mitigation. In addition to the aforementioned fact that stopping animal exploitation can add remarkable speed to bursting the meat bubble and hence mitigation (switching to a vegan diet does not require "infrastructure" being laid down first) at a much more profound level, rethinking human–animal relations creates an organismic change with a multiplying effect for climate mitigation. What I refer to is the effect of quitting neoliberal empathy- and compassion-inhibitors. The realization of neoliberal control techniques gives individuals autonomy. When individuals reject neoliberal luring, coaxing, and nudging and choose to become reconnected to their deeper emotions, such as empathy and compassion, the consequences cannot be limited to single issues: It is not possible for a person to be able to feel the suffering that another being has to endure if the source of the suffering is from x, but not if it is from y or z. The documentary Live and Let Live shows that many animal rights activists became aware of and vigilant to other forms of exploitation including in the areas of environment and labor. By feeling the pain that animals feel, the activists' senses became even more sharpened. Not noticing the suffering caused by other forms of exploitation became less and less possible.[15]

Compassion is the reason that there is still hope for climate mitigation. For whatever reason, animals are good at helping humans to regain humanity and expand their compassion, and this is good news for climate action. Neoliberals, particularly "rationalists," will condemn the faith in compassion as being touchy-feely, "irrational," or "unscientific." Charles Darwin would disagree. Herbert Spencer turned Darwin's theory on its head and twisted the meaning of "survival of the fittest" and "natural selection" in constructing his social Darwinism (Spencer, 1866: 444–5). In *Descent of Man*, Darwin talked about "social virtues" like "sympathy,[16] fidelity, and courage." He stated "Such social qualities, the paramount importance of which to the lower animals is disputed by no one, were no doubt acquired by the progenitors of man in a similar manner, namely, through natural selection, aided by inherited habit" (Darwin, 1922 [1871]: 199). What Darwin meant by "natural selection," therefore, is antithetical to what Spencer meant in his social Darwinism.[17] Darwin stated:

> When two tribes of primeval man... came into competition, if... the one tribe included a great number of courageous, sympathetic and faithful members, who were always ready to warn each other of danger, to aid and defend each other, this tribe would succeed better and conquer the other. (Darwin, 1922 [1871]: 199)

Darwin explained that the reason that sympathy, courage, and faithfulness would spread is because:

> each man would soon learn that if he aided his fellow-men, he would commonly receive aid in return. From this low motive he might acquire the habit of aiding his fellows; and the habit of performing benevolent actions certainly strengthens the feeling of sympathy which gives the first impulse to benevolent actions... But another and much more powerful stimulus to the development of the social virtues, is afforded by the praise and the blame of our fellow-men. (Darwin, 1922 [1871]: 201)

Spencer's attribution of social Darwinism to Darwin fits right into George Orwell's novel[18] and is explanatory of neoliberals' malicious emphasis on "individualism" and "competitiveness." The more that societies praise virtues such as sympathy and other benevolent actions, the more they will value cooperation, what is public, or what belongs to the commons, and the less there will be left for the greedy to grasp and keep as their private properties.

Contrary to social Darwinism, Darwin saw every reason for individuals to extend empathy and compassion beyond their immediate surroundings. The following quotation links human, animal, climate change, and neoliberalism.

> As man advances in civilization, and small tribes are united into larger communities, the simplest reason would tell each individual that he ought to extend his social instincts and sympathies to all the members of the same nation, though personally unknown to him. This point being once reached, there is only an artificial barrier to prevent his sympathies extending to the men of all nations and races... Sympathy beyond the confines of man, that is, humanity to the lower animals, seems to be one of the latest moral acquisitions. It is apparently unfelt by savages, except towards their pets. How little the old Romans knew of it is shewn by their abhorrent gladiatorial exhibitions... This virtue, one of the noblest with which man is endowed, seems to arise incidentally from our sympathies becoming more tender and more widely diffused, until they are extended to all sentient beings. (Darwin, 1922 [1871]: 188)

Darwin's words, written one and a half centuries ago, combined with evidence poignantly presented by aforementioned animal behavioralists demonstrating that "there is more thought and feeling in animals than humans have ever imagined," that they "are highly sentient and sometimes even virtuous," and that they "experience pain, pleasure, and emotions, and their lives have meaning beyond any utilitarian value that humans may place on them" (Balcombe, 2011: 4; 13), indicate that humans have made a terrible mistake treating animals the way that we do. Not only do animals feel, but we also care. Today, neoliberals, the "savages" and "old Romans" of our time, are at the forefront of making sure that this mistake, together with other ones, such as burning fossil fuels, do not get corrected. As Darwin's observation that "communities, which included the greatest number of the most sympathetic numbers, would flourish best" (Darwin, 1922 [1871]: 163), implies, the survival of the human and many other species hinges on the ability of societies to replace the neoliberal zombie democracy with inclusive democracy.

Notes

1 As a method of boycotting climate offenders, the divestment movement first appeared in the U.S. in 2000 when Ozone Action urged investors to divest from U.S. companies that participated in the climate-denial Global Climate Coalition

(GCC). The divestment campaign was considered a success, as the GCC dissolved in 2002 (Mayes *et al.*: 135).

2 The first fossil fuel divestment campaign began at Swarthmore College over a year before 350.org became involved. Throughout 2011, in-depth planning conversations took place among students at Swarthmore College and the University of North Carolina at Chapel Hill and organizations including the Energy Action Coalition, As You Sow, the Sierra Student Coalition, the Sustainable Endowments Institute, the California Student Sustainability Coalition, and the Responsible Endowments Coalition (Bourqui, 2012).

3 Defunding entails pressuring banks to stop providing loans financing new fossil fuel projects such as new pipelines or fracking drill rigs. See https://gofossilfree. org/not-a-penny-more/.

4 The cancer agency of the World Health Organization-the International Agency for Research on Cancer-classified beef, lamb, and pork as carcinogenic when eaten in processed form and as probably carcinogenic if eaten unprocessed (Bouvard et al., 2015).

5 A study by Springmann, Clark *et al.* covering 149 world regions estimate that 2.4 million deaths in 2020 will be attributable to red and processed meat consumption, and health-related costs to society will amount to some $285 billion (2018: 1, 11).

6 Stephen Gill uses Gramsci's concept of "organic crisis" to describe the current world. Environmental degradation resulted from animal exploitation can easily fit into this concept.

7 https://sentientmedia.org/how-many-animals-are-killed-for-food-every-day/.

8 The Kantian deontological perspective stipulates that rights and duties tend to override utilitarian considerations (Greene et al., 2008: 1145).

9 Nibert points out that "The oppression of humans and other animals developed in tandem, each fueling the other" (Nibert, 2002: 50).

10 Francione described how, by selling animal welfare ideas, corporate charities stopped the momentum of a genuine animal rights movement that came very close to making a real impact in the mid-1990s (Francione, 2018: 17–20).

11 Consider the rationale implicit in most neoliberal justification of hazardous products: While pollution may cut our lives short, the prosperity that necessitates pollution has enriched our lives. Unhealthy foods may cut our lives short, the savoring of such heavenly foods is itself priceless experience that nobody has the right to take away.

12 This 2010 ruling concerned the restrictions of political spending on electoral advocacy imposed by the Bipartisan Campaign Reform Act. The U.S. Supreme Court held that political spending is a form of free speech protected by the First Amendment and that corporations, unions, and profitable organizations are entities entitled to the First Amendment protection. See https://supreme. justia.com/cases/federal/us/558/310/.

13 It is critical to recognize that the exclusion of the wellbeing of some humans and animals from consideration did not result from prejudice; rather, prejudice was socially constructed to legitimate oppression that served the interests of a powerful elite (Nibert, 2002: 31, 52). Not only did the oppression of animals facilitate that of humans and vice versa, but the devices used on controlling animals were also later used on controlling humans. "Barbed wire was first introduced to control the movements of other animals...[and later] became the central element in the architecture of the death camps" (Nibert, 2002: 73). "Electric cattle prods were used in the 1960s by police officers in some southern cities against African Americans at civil rights demonstrations" (Nibert, 2002: 76).

14 Consider the rhetoric of economic growth measured by GDP.

15 Francione points out that the animal rights movement he and Regan led had from the very beginning took human rights and animal rights issues as inseparable. The conferences they held "brought together farm workers and labor, civil rights advocates, feminists, gay/lesbian advocates... to discuss the common issues of rights and justice" (Francione, 2018: 15).

16 Citing Adam Smith and Alexander Bain, Darwin noted that the basis of the "all-important emotion of sympathy... lies in our strong retentiveness of former states of pain or pleasure. Hence, the sight of another person enduring hunger, cold, fatigue revives in us some recollection of these states, which are painful even in ideas. We are thus impelled to relieve the sufferings of another, in order that our own painful feelings may be at the same time relieved. In like manner we are led to participate in the pleasures of others... The mere sight of suffering... would suffice to call up in us vivid recollections and associations" (Darwin, 1922 [1871]: 162–3).

17 Like Bentham, Spencer has a prominent place in Polanyi's discussion of enclosure and the market force contributing to the double movement. As a strong supporter of *laissez-faire*, Spencer was "horror struck" by "restrictive legislation" such as the "Mines Act making it penal to employ boys under twelve not attending schools and unable to read or write," and the "Chimney-Sweeper's Act, to prevent the torture and eventual death of children set to sweep too narrow slots," and the "Public Libraries Act, giving local powers by which a majority can tax a minority for their books." "Spencer adduced [such legislation] as irrefutable evidence of an anti-liberal conspiracy" (Polanyi, 2001 [1944]: 152).

18 Where, for instance, "war is peace, freedom is slavery, and ignorance is strength" (Orwell, 1987 [1949]).

References

Alexander, Peter, Calum Brown, Almut Arneth, John Finnigan, Dominic Moran, and Mark D. A. Rounsevell. (2017). "Losses, Inefficiencies and Waste in the Global Food System," *Agricultural Systems*, Vol.153: 190–200.

Alix-Garcia, Jennifer and Holly K. Gibbs. (2017). "Forest Conservation Effects of Brazil's Zero Deforestation Cattle Agreements undermined by Leakage," *Global Environmental Change*. Vol.47: 201–217.

Ansar, Atif, Ben Caldecott and James Tilbury. (2013). Stranded Assets and the Fossil Fuel Divestment Campaign: What Does Divestment Mean for the Valuation of Fossil Fuel Assets? University of Oxford Stranded Assets Program.

Arabella Advisors. (2018). The Global Fossil Fuel Divestment and Clean Energy Investment Movement.

Balcombe, Jonathan. (2006). *Pleasurable Kingdom: Animals and the Nature of Feeling Good*. London: Palgrave Macmillan.

Balcombe, Jonathan. (2011). *Second Nature: The Inner Lives of Animals*. New York: St. Martin's Griffin.

Balcombe, Jonathan. (2017). *What A Fish Knows*. New York: *Sci Am* and Farrar, Straus and Giroux.

Bekoff, Marc. (2008). *The Emotional Lives of Animals: A Leading Scientist Explores Animal Joy, Sorrow, and Empathy*. Novato, CA: New World Library.

Bigelow, Bill. (2013). "Teaching the Terrifying Math of Climate Change," CommonDreams.org. 20 December.

Bourqui, Martin. (2012). "Divestment Proponents: Our Math is Sound," *Huffington Post*. 23 December.

Bouvard, Véronique, Dana Loomis, Kathryn Z. Guyton, Yann Grosse, Fatiha El Ghissassi, Lamia Benbrahim-Tallaa, Neela Guha, Heidi Mattock, Kurt Straif, and International Agency for Research on Cancer Monograph Working Group. (2015). "Carcinogenicity of Consumption of Red and Processed Meat," *The Lancet Oncology*. Vol. 16(16): 1599–1600.

Darwin, Charles. (1922 [1871]). *Descent of Man And Selection in Relation to Sex*. London: John Murray.

De Waal, Frans. (1997). *Good Natured: The Origins of Right and Wrong in Humans and Other Animals*. Cambridge, MA: Harvard University Press.

De Waal, Frans. (2017). *Are We Smart Enough to Know How Smart Animals Are?* New York: W. W. Norton & Company.

De Waal, Frans. (2019). *Mama's Last Hug: Animal Emotions and What They Tell Us About Ourselves*. New York: W. W. Norton & Company.

Donaldson, Sue and Will Kymlicka. (2011). *Zoopolis—A Political Theory of Animal Rights*. Oxford: Oxford University Press.

Earthlings. (2005). Documentary film directed by Shaun Monson. Malibu: Nation Earth.

Eckelman, Matthew J. and Jodi Sherman. (2016). "Environmental Impacts of the U.S. Health Care System and Effects on Public Health," *PLoS One*. 11(6): e0157014. https://doi.org/10.1371/journal.pone.0157014.

Edwards, Morgan R. and Jessika E. Trancik. (2014). "Climate Impacts of Energy Technologies Depend on Emissions Timing," *Nature Climate Change*. Vol. 4. 25 April. DOI: 10.1038/NCLIMATE2204.

Evich, Helena Bottemiller. (2016). "Meat Industry Wins Round in War over Federal Nutrition Advice." *Politico*. 7 January.

FAIRR. (2017). *The Livestock Levy*. 11 December.

Food and Agriculture Organization of the United Nations. (2006). *Livestock's Long Shadow—Environmental Issues and Options*. Rome: Food and Agricultural Organization.

Food and Agriculture Organization of the United Nations. (2009). *The State of Food and Agriculture—Livestock In the Balance*. Rome: Food and Agricultural Organization.

Francione, Gary. (1995). *Animals, Property, and the Law*. Philadelphia, PA: Temple University Press.

Francione, Gary. (2018). "Reflections on Tom Regan and the Animal Rights Movement That Once Was," *Between the Species*. 21(1): 1–41.

Galbraith, John K. (2007 [1967]). *The New Industrial State*. Princeton, NJ: Princeton University Press.

Geman, Ben. (2017). "Shell CEO Warns of 'Disappearing' Public Patience on Carbon Emissions," *Axios*. 10 March.

Gerber, P.J., H. Steinfeld, B. Henderson, A. Mottet, C. Opio, J. Dijkman, A. Falcucci, and G. Tempio. (2013). *Tackling climate change through livestock–A global assessment of emissions and mitigation opportunities*. Rome: Food and Agriculture Organization of the United Nations.

Gibbons, Chip. (2017). Ag-Gag Across America—Corporate-Backed Attacks on Activists and Whistleblowers. Report published by the Center for Constitutional Rights and Defending Rights & Dissent.

Gill, Stephen. (2015). Global Organic Crisis and Geopolitics. *AnalyzeGreece!*. 5 June.

Grace, Peter and Louise Barton. (2014). "Meet N2O, The Greenhouse Gas 300 Times Worse Than CO2," *Conversation*. 9 December.

Greene, Joshua D., Sylvia A. Morelli, Kelly Lowenberg, Leigh E. Nystrom, and Jonathan D. Cohen. (2008). "Cognitive Load Selectively Interferes with Utilitarian Moral Judgment," *Cognition*. 107: 1144–1154.

Greenwald, Glenn. (2018). "Six Animal Rights Activists Charged with Felonies for Investigation and Rescue That Led to Punishment of a Utah Turkey Farm," *The Intercept*. 4 May.

Griscom, B. W. *et al.* (2017). "Natural climate solutions," *Proceedings of the National Academy of Sciences*, 114(44), 11645–11650.

Hedges, Chris. (2013). "The Business of Mass Incarceration," *Truthdig*. 29 July.

Heid, Markham. (2016). "Experts Say Lobbying Skewed the U.S. Dietary Guidelines," *Time*. 8 January.

Henriques, Irene and Perry Sadorsky. (2017). "Investor Implications of Divesting from Fossil Fuels," *Global Finance Journal*. Vol.38: 30–44.

Heuer, Jason. (2016). "No More Exposés in North Carolina," *The New York Times*. 1 February.

Hudson, Michael. (2017). *J Is for Junk Economics—A Guide to Reality in an Age of Deception*. Glashütte: ISLET-Verlag.

IPCC. (1990). *Climate Change—The IPCC Scientific Assessment*.

IPCC. (2018). *Special Report: Global Warming of 1.5°C*. October.

Jacobsen, Stefan Gaarsmand. (2018). "Climate Justice as anti-Corporate Economic Mobilization," In Stefan Gaarsmand Jacobsen (ed.), *Climate Justice and the Economy—Social Mobilization, Knowledge and the Political*, 3–31.

Kearns, Cristin E., Laura A. Schmidt and Stanton A. Glantz. (2016). "Sugar Industry and Coronary Heart Disease Research: A Historical Analysis of Internal Industry Documents," *JAMA Internal Medicine*. 176(11): 1680–1685.

Klein, Naomi. (2015). Naomi Klein on the Power of Fossil Fuel Divestment. Interview with *Grist*. 11 February.

Lenferna, Georges Alexandre. (2018). "Divestment as Climate Justice—Weighing The Power of the Fossil Fuel Divestment Movement," In Stefan Gaarsmand Jacobsen (ed.), *Climate Justice and the Economy—Social Mobilization, Knowledge and the Political*, 84–109.

Mayes, Robyn, Carol Richards, and Michael Woods. (2017). "(Re)assembling Neoliberal Logics in the Service of Climate Justice: Fuzziness and Perverse Consequences in the Fossil Fuel Divestment Assemblage." In Vaughan Higgins and Wendy Larner (eds.), *Assembling Neoliberalism—Expertise, Practices, Subjects*. New York: Palgrave Macmillan, 131–149.

McKibben, Bill. (2012). "Global Warming's Terrifying New Math," *Rolling Stone*. 2 August.

Montgomery, Sy. (2016). *The Soul of an Octopus: A Surprising Exploration into the Wonder of Consciousness*. New York: Atria Books.

Nibert, David. (2002). *Animal Rights Human Rights—Entanglements of Oppression and Liberation*. London: Rowman & Littlefield Publishers.

Nozick, Robert. (1974). *Anarchy, State, and Utopia*. New York: Basic Books.

Nussbaum, Martha. (2004). "Beyond 'Compassion and Humanity': Justice for Nonhuman Animals." In Martha C. Nussbaum and Cass R. Sunstein (eds.),

Animal Rights: Current Debates and New Directions. Oxford: Oxford University Press: 299–320.

Opio, C. *et al*. (2013): *Greenhouse Gas Emissions from Ruminant Supply Chains—A Global Life Cycle Assessment*. Rome: Food and Agriculture Organization of the United Nations.

Oppel, Richard. (2013). "Taping A Farm Cruelty Is Becoming A Crime," *The New York Times*. 6 April.

Orwell, George. (1987 [1949]). *Nineteen Eighty-Four*. London: Secker & Warburg.

Parenti, Christian. (2013). "Problems with the Math: Is 350's Carbon Divestment Campaign Complete?" *Huffington Post*. 29 January.

Regan, Tom. (2004 [1983]). *The Case for Animal Rights*. Berkeley: University of California Press.

Riahi K. *et al*. (2017). "The Shared Socioeconomic Pathways and Their Energy, Land Use, and Greenhouse Gas Emissions Implications: An Overview," *Global Environment Change*. 42: 153–168.

Safina, Carl. (2015). *Beyond Words: What Animals Think and Feel*. New York: Henry Holt & Company Inc.

Sanzillo, Tom, Kathy Hipple, Clark Williams-Derry. (2018). The Financial Case for Fossil Fuel Divestment. Sightline Institute and Institute for Energy Economics and Financial Analysis.

Schifeling, Todd and Andrew J. Hoffman. (2017). "How Bill McKibben's Radical Idea of Fossil-Fuel Divestment Transformed the Climate Debate," *The Conversation*. 12 December.

Schinkel, Anders. (2008). "Martha Nussbaum on Animal Rights," *Ethics & The Environment*. 13(1): 41–68.

Shell. (2018). Strategic Report, Shell Annual Report and Form 20-F 2017. 14 March.

Singer, Peter. (2009 [1975]). *Animal Liberation*. New York: Harper Collins.

Spencer, Herbert. (1866). *The Principles of Biology*. New York: D. Appleton & Company.

Springmann, Marco, M. Clark, D. Mason-D'Croz, K. Wiebe, B. L. Bodirsky, L. Lassaletta, W. Vries, S. J. Vermeulen, M. Herrero, K. M. Carlson, M. Jonell, M. Troell, F. DeClerck, L. J. Gordon, R. Zurayk, P. Scarborough, M. Rayner, B. Loken, J. Fanzo, H. Charles J. Godfray, D. Tilman, J. Rockström, and W. Willett. (2018). "Options for Keeping the Food System within Environmental Limits," *Nature*. 10 October.https://doi.org/10.1038/s41586-018-0594-0.

Springmann, Marco, Daniel Mason-D'Croz, Sherman Robinson, Keith Wiebe, H. Charles J.Godfray, Mike Rayner, and Peter Scarborough. (2018). "Health-Motivated Taxes on Red and Processed Meat: A Modelling Study on Optimal Tax Levels and Associated Health Impacts," *PLoS One*. 13(11): e0204139.

Stephenson, Wen. (2012). "Cue the Math: McKibben's Roadshow Takes Aim at Big Oil," *Grist*. 18 October.

Thornton, Philip, M. Herrero and P. Ericksen. (2011). "Livestock and Climate Change." *Livestock Xchange*. International Livestock Research Institute. November.

Trinks, Arjan, B. Scholtens, M. Mulder, and L. Dam. (2018). "Fossil Fuel Divestment and Portfolio Performance," *Ecological Economics*. Vol. 146: 740–748.

Veiga, J. B., J. F. Tourrand, R. Poccard-Chapuis and M.G. Piketty. Cattle Ranching in The Amazon Rainforest. Paper Presented to the XII World Forestry Congress. Québec City, Canada.

Watkins, John and James Seidelman. (2019). The Last Gasp of Neoliberalism. Paper presented before the Association for Evolutionary Economics and Allied Social Science Association. Atlanta, Georgia. 4 January.

Wolf, Julie, G. R. Asrar and T. O. West. (2017). "Revised Methane Emissions Factors and Spatially Distributed Annual Carbon Fluxes for Global Livestock," *Carbon Balance and Management*. 12:16. DOI 10.1186/s13021–13017–0084-y.

World Meterorological Organization. (2018). *WMO Greenhouse Gas Bulletin*. 22 November, No. 14.

Zickfeld, Kirsten, S. Solomon and D. M. Gilford. (2017). "Centuries of Thermal Sea-Level Rise Due to Anthropogenic Emissions of Short-Lived Greenhouse Gases," Proceedings of National Academy of Science of the U.S.A. 114(4): 657–662.

8 Epilogue

Imagine a world where the distribution of wealth is exactly the same as ours: 26 individuals own as much as the poorer half of the entire world (Oxfam, 2019). In that world, however, the money is just *handed* to the wealthy by the rest of society through a bizarre tax system, instead of being channeled through the manufacturing, sale, and consumption of commodities harmful to the health of the public, the normal functioning of the economic system, or the environment. This imagined world would be so much better than the one that we live in. Few or no people would have cancer, heart disease, diabetes, obesity, respiratory disease, or chemical-caused allergies. The environment would be well preserved. There would be no burning of fossil fuels and rising sea levels. Wild animals would be left alone in their sovereign kingdom. Domesticated animals would live alongside humans not as food or clothes but as family and friends. No people, or very few people, would be forced to become refugees fleeing from wars, which would be rare, or climate-related disasters, which also would be rare. It is impossible that anyone, given a choice, would choose our world over the imagined one where people are healthy and the environment is beautiful.

That world, of course, is immensely impractical. The bizarre tax system would have absolutely no reason to be there. This is why the fairy tale of one big self-regulating market had to be invented. The discourse of the self-regulating market performs the function of the bizarre tax system. It fends off regulations protecting public interests and ensures the upward redistribution of wealth. Societies have to be twisted and torn to fit the fairy tale. The self-regulating market is presented as a sophisticated machine, the operation of which needs to be left completely in the hands of the specially trained mechanics called economists. Laymen can have no say about this delicate machine, let alone touch it or interfere with it. Keynes noticed the problem with this machine, which lacked the ability to observe or to care whether people lived well or miserably, or even remained alive at all. The Keynesian revolution made regulating the callous market sane. Many intractable problems of our time, climate change above all, can be traced to the success of business-funded efforts undoing the damage that Keynes did to the fairy tale.

While it is a commonplace to portray "humans" as a collective as responsible for problems such as climate change, owing to their greed and disrespect of nature, such narratives grotesquely shift the main responsibility from a small group of powerful people steering and arranging the global socioeconomic order to the rest of society. The rest of society is not responsibility-free, of course. Their most urgent duty at the current juncture is to see through neoliberal lies and find smart ways to bypass, undermine, and eventually overthrow the planet-destroying SCAMD structure.

Reference

Oxfam. (2019). *Public Good or Private Wealth?* January.

Index